WELCOME

TO THE 5TH DIMENSION

LIONEL INGRAM

AN EXPOSÉ OF THE HIDDEN HISTORY OF THE COSMOS

ISBN
978-1-5437-4920-5 (sc)
978-1-5437-4919-9 (e)

Library of Congress Control Number: 2019934156

First Edition 2018
Edited by David Kaplan
www.freelanceeditors.co.za

Print information available on the last page.

To order additional copies of this book, contact
Toll Free 800 101 2657 (Singapore)
Toll Free 1 800 81 7340 (Malaysia)
www.partridgepublishing.com/singapore
orders.singapore@partridgepublishing.com

03/27/2019

PARTRIDGE

WELCOME TO THE 5TH DIMENSION
A BOOK BY LIONEL INGRAM

An Exposé of the Hidden History of the Cosmos

First Edition 2018

THIS BOOK
IS DEDICATED TO EVERY PERSON
ABLE TO PERCEIVE AND APPRECIATE
THIS UNDENIABLE TRUTH.

CONTENTS

INTRODUCTION

This book provides a clue to unlocking perhaps the greatest mystery ever anticipated. It presents evidence of a history (hidden in plain sight) that will alter our perception and understanding of the universe, showing how we are embedded in a dimension that is based upon a fractal.

Fractals are typically self-similar patterns, where 'self-similar' means they are the same whether seen from near or from far. Fractals may either be the same at every scale, or else almost the same at different scales. The definition of a fractal goes beyond self-similarity per se to exclude trivial self-similarity and include the idea of a detailed pattern repeating itself.

The Romanesco broccoli (see below) is one of the best examples indicating a typical fractal pattern in nature. Notice the shape and form, zoomed both in or out within this macro view consisting of a repeat pattern of cone-shaped heads.

Photo provided by FeaturePics. All Right Reserved

Fractals are found throughout nature and we are exposed to fractals daily without realising it. With this in mind, it is worth noting how similar the world of the atom is, with its orbiting electrons and protons mimicking the stars and orbiting exoplanets.

The famous Fibonacci sequence has captivated mathematicians, artists, designers and scientists for centuries. Also known as the **Golden Ratio**, 1.618, this typically governs fractal designs.

Other examples created by using mathematics have produced beautiful repeating patterns incorporating the Golden Ratio. This ratio appears in many patterns in nature, including the spiral arrangement seen in flowers, shells and even galaxies.

This spiral may be the means to move from one level to another, such as from a bigger part of a fractal to a smaller part in a never-ending continuum.

When observing spiral galaxies in the universe, we can immediately notice how these galaxies are governed by this ratio. The Milky Way (our galaxy) is a spiral galaxy.

Our knowledge of the known universe is limited to our 3-dimensional perception of space, with time, the 4th dimension, governing the physical attributes of all matter, such as aging and decay (Newton's Third Law of Thermodynamics). Time, as we know it, is what prevents our ability to reason beyond the speed of light.

The speed of light itself is determined by time, i.e. 299,792 km per **second** and, in our space-time continuum, it is said to be a constant. The magnitude of most of the observations revealed in this book will certainly provoke a huge question concerning the **speed of light**, and how we understand it as a constant within the 4th dimension of our current time-space.

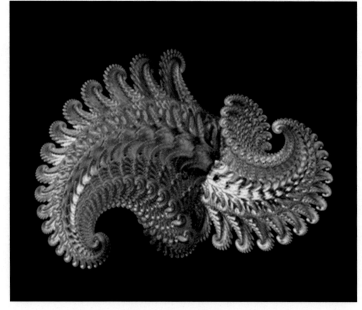

Above: An Artist impression of a Fractal. Created by Mr. Jock Cooper
and displayed here with his kind permission.

This book uncovers a *world* never contemplated before. It shows how things once considered myth and fable were in fact a reality and are discernible through a technique (utilising filters) that is applied when observing the Earth, the planets of our solar system and the stars making up the galaxies.

From this perspective, the book provides the reader with credible evidence of the prior existence of various life forms, having existed in the past, which now provide the very substance of matter forming the stars and planets in our known universe. These life forms ranged hugely in size, some of mind-boggling proportions, as will be seen in the pages that follow. Due to these observations, the speed of light, as we currently understand it, comes into question. It would seem logical to assume that everything was scaled proportionately to accommodate whatever size the life forms displayed herein may have been. This also means that the speed of light itself was accommodating the larger life forms proportionately in whatever dimension they existed in. If this hypothesis is true, then the creation of the universe may have been a separate event beyond the scope of the Big Bang theory.

For example, a single day (24 hours) in the dimension scaled up to match the largest entity found on Earth thus far (see pages 90 and 91) would cover an equivalent period amounting to 47 years in our time span here on this planet. This entity most likely existed long before the solar system, as we now know it was formed. This can be logically deduced when realising the size of the entity in question. If it could sprint like an athlete, e.g. a 100-meter dash in its dimension, it would appear from our current time-space perspective that it would be running at approximately 56% of the speed of light. What, then was the speed of light in its dimension? A reasonable conclusion might be that the speed of light, with its specific constant is scaled proportionately to suit the particular time-space it occupies. All organic matter (flora and fauna) consists of coded DNA, shaping every kind of life form imaginable. Here, evidence is presented revealing life forms so large they could never have existed in our solar system as we know it today. But it does not end here. Concrete evidence is presented showing that even larger life forms existed *before galaxies were formed*!

Now, this may sound like a bold statement, however, when all the evidence presented in this book is taken seriously, it will be proven true, and the proof thereof is audaciously displayed herein. It is also entirely feasible that the evidence presented is a visual representation of the so-called 'Akashic Records'. According to theosophy and anthroposophy, the Akashic Records are a compendium of all human events, thoughts, words, emotions and intent ever to have occurred in the past, or that will occur in the present or future. All of history, from the beginning of time, to the end of time, is encoded in the DNA.

What then, precisely is the DNA Code? If we consider the fact that the Code itself is entirely the reason why everything exists and is itself comprised of the atoms of the universe, then we need to think a little deeper. Here is the source of all beginnings encaptured within a fractal, encoded in the DNA molecule.

Accepting then that fractal time could be governed within the DNA, with each interval representing a different epoch, we can draw a conclusion. Given the afore-mentioned, contained herein is supporting evidence presenting a variety of various-sized life forms that coexisted during certain time cycles. This is also supported by the observation of the systematic diminishing of larger life forms into smaller components over time. Clearly, this assumes an arranged fractal within which cycles are governed, each cycle seemingly ending in a sudden cataclysmic event. This sudden event leaves all organic matter in a state of frozen inertness (a type of crystallisation, not necessarily a cold freeze), which eventually becomes rock. A new cycle then begins, a rebirth of all organic matter assuming a smaller size based on the Golden Ratio of 1.618. Out of the dust of the Earth we are created and back to dust we will go. So too is the cycle of life, established in accordance to the Fibonacci sequence, and perhaps expounding the mechanics of the entire universe since its beginning.

While viewing the evidence, the reader may immediately suspect 'photoshopping' as the responsible medium for easily creating any of the features seen within these pages. However, this book contains no 'photoshopping' in the true sense, but rather comprises a technique whereby detail hidden to the eye may readily be uncovered, much like an archaeologist or palaeontologist, after identifying certain features, will meticulously remove any debris that may be hiding the detail. What you will witness here is archaeology / palaeontology on a grand scale, but without using picks or shovels, breaking rock or removing soil. It is rather the utilisation of suitable digital tools to remove the noise, revealing the underlying image. Many of the features are embedded with other things, but despite this, much can still be individually identified by using simple techniques to bring out precise detail, as can be appreciated in the image below (this is **not** a drawing).

UNCOVERING THE REALITY OF MYTH

Warning: Some of the images in this book may be disturbing to closed-minded readers.

CHAPTER 1

THE
RED PLANET

THE BEGINNING OF A JOURNEY...

It was around the end of December 2012 while driving to work and listening to a popular Johannesburg Radio Station (702) morning talk show that the presenter stated some recent developments declared by NASA following the landing of the Curiosity Rover on the surface of Mars, in August of the same year. It was announced that NASA was preparing an official report to be published within two weeks about a discovery made on the planet, quoted as 'One for the history books'.

Having a keen interest in space travel and exploration, I found this announcement fascinating and waited with bated breath for this 'historical' event to be revealed to the world. However, when the time came around mid-January 2013 for the so-called disclosure, NASA changed the entire story with a rather weak statement that their data was flawed, and they therefore retracted their previous statement that was supposedly going to shake the world. This left me and perhaps many others somewhat puzzled. Why would an organisation such as NASA want to disclose something to the media without firstly verifying or validating data **before** making such an announcement, especially if it was going to 'change our history books'? And why did Pope Benedict XVI resign his position shortly afterwards in February 2013??

In March of 2013, NASA set up an interactive website (http://mars.nasa.gov/multimedia/ interactives/billionpixel/) for the general public to access and to participate in the current programme, providing among other things a panoramic viewer showing a billion-pixel view of the surface of Mars captured by the cameras operating from Curiosity Rover. The panoramic viewer also provided the facility to zoom or pan in and out, allowing the user to examine various features around the location where the rover was situated, in the so-called Gale Crater on Mars.

I visited this site and made an astounding observation. While panning and zooming in and out, looking at the skyline, landscape, 'rocks' etc., I found a human-looking hand. Upon closer examination I was convinced that I was not looking at a 'rock'. This was followed by a flood of findings over the next few weeks of various things of which I began collecting screenshots. This book displays only a small part of a huge collection that is available.

NASA had apparently discovered something on Mars, and yes indeed, it **will** change our history books!

Since NASA's first close-up picture of Mars in 1965, and up to the present day, spacecraft voyages to the Red Planet have revealed a world surprisingly familiar, yet sufficiently different to confront our perceptions of what really existed on this planet.

Water, as we know, is the key to life, and the earlier Mars missions, according to NASA (2001 Mars Odyssey, Mars Exploration Rovers, Mars Reconnaissance Orbiter and Mars Phoenix Lander), were designed to make discoveries under the Mars Exploration Program science theme of 'Follow the Water', where progressive discoveries relating to evidence of past and / or present water in the geologic record would make it possible to take the next steps towards finding evidence of life itself. It is now well established that Mars had abundant water in the past.

The question as to whether Mars provided similar environmental conditions for life in the distant past has been answered by NASA. A more relevant question now, however, is why NASA chose not to reveal the history-changing revelation they promised us for early 2013. Could it be possible that the truth is considered too much to bear? So much so that it resulted in the resignation of Pope Benedict XVI? This idea could be supported by the fact that NASA may have consulted immediately with the Vatican following what was discovered on Mars in late 2012. The images presented in this chapter explains the reason.

All the images presented in this section were obtained by thorough searching (panning and zooming) around the debris seen scattered all over the area in the locality of the Gale Crator, as viewed from the Curiosity Rover. All the screenshots were obtained off the original site set-up (2013) provided to the public from the website noted above. However, at this present time, NASA have since upgraded the site to inexplicably exclude the zoom capability that the original site provided. This now prohibits the viewer from scrutinising detail as was previously possible. Why are they limiting access to detail and what are they hiding?

Unfortunately, I cannot provide the reader with specific coordinates.The images are simply 'screenshots' taken by utilising the printscreen option usually available on most computers. In some cases, I have included a screenshot showing the exact location (the hand-icon) where the reader could have located some of these images as they appeared on the older system.

NASA had originally provided the public the means to observe exactly what is on Mars. The onboard hi-resolution camera delivered the opportunity for anyone to make observations, and to draw their own conclusions. I believe it was NASA's original intention to make this disclosure at the outset of their announcement. However, the fact that NASA has now withdrawn the capability of close examination previously provided prevents the individual from drawing any (possibly inconvenient) conclusions and raises a myriad of questions.

The pictures presented in this book are but a sample of hundreds of features that were secured from this site. Contrast and equalising colour, shadow and light is all that it takes to reveal the reality. My fellow kindred, we are mature enough to rely on our own observations and do not require the opinions of scientists or experts to confirm what we are looking at. The overwelming evidence presented should jolt any person with an average IQ to realise the reality before us.

Here evidence is presented that life had once existed on Mars, albeit not what we might expect or believe. Each picture presented herein shows the raw image, as I was able to capture it as a screenshot on the interactive billion-pixel site. Alongside the same image I have added contrast and / or equalised colour, shadow and light to accentuate the view.

Each picture has been named to identify the various characteristics noted.

Below is a screenshot (taken in 2013) from NASA's Interactive Billion-Pixel site showing a view of Mars from Curiosity Rover.

Note that the following screenshots are not for the fainthearted and some readers may find these images disturbing.

The image below indicated with the hand-icon in the centre marks the exact location where the following features could be found.

Zooming into this area and carefully examining the detail revealed much more than what I expected to be mere 'rocks'. The mountain seen in the distance is called Mt Sharp standing about five kilometres high, located more or less at the centre of the crater.

The image below shows what the investigator would have found. Noted here is a hand and wrist with a sleeve partially protruding out of the rubble, a bearded face of a small impish man (only the head with beard is noticeable) and a dog-like creature, all grouped closely together.

Hand, Dog and Geni

| **Raw Image** | **Enhanced image** |

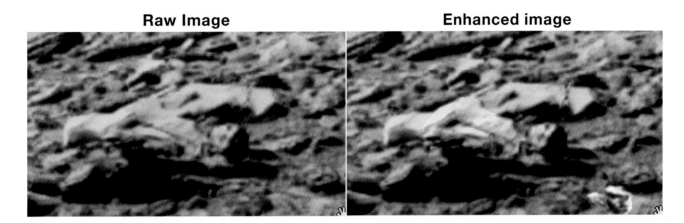

The little black-bearded face appears to be a type of impish, geni-looking being with a visible horn and a pointed ear.

The dog's head, facing the camera, is tilted slightly to the left, showing one eye, its nose and glinting teeth and appears somewhat poodle-like.

The hand appears to have a strap attached to the wrist, possibly a leash leading to the dog. Many other features can also be noticed, not all of which are highlighted.

All of these features were found in a single group, each of them bearing characteristics that define specific attributes that are clearly recognisable. These are not shadows casting a trick on the eye, and anyone so naïve to believe that this is an illusion simply cannot perceive reality.

The zoomed-in view below shows the detail of the hand. Note what appears to be nail polish reflecting off the thumbnail along with a leash strapped around the wrist leading to the dog.

The detail seen in all of the screenshots assume the same attributes, revealing the stark reality of what actually existed on Mars. Nothing is drawn or added apart from contrast and colour adjustments.

This picture alone presents all the shocking evidence required to confirm that life had once existed on the Red Planet before everything was totally obliterated by some enormous explosion or possibly a meteorite impact that formed the Gale Crater, the current location where the Curiosity Rover resides.

Little Geni and the Poodle-like Dog.

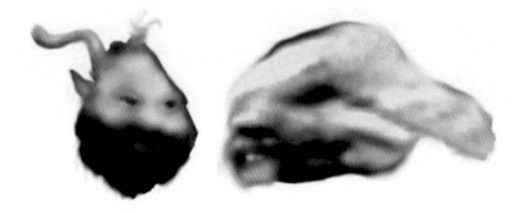

Many other objects and features can be noted scattered around this specific location. Careful observation is required to fully realise the extent of the detail.

Below, only a few meters away (proportionately within the interactive viewer), at about a 45° angle above the hand-icon as seen on page 4, a zoomed-in view reveals some dark-looking characters. The detail here reveals a very different expectation of the type of life forms that existed on this planet.

Dark Associates

Raw Image   **Enhanced image**

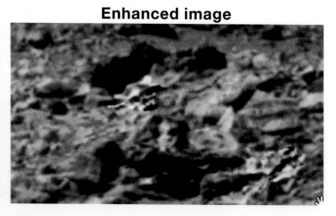

Among other things, the detail in this screenshot reveals a scull with strange protrusions and other beings of a dark appearance, one of which appears to be holding books and scrolls. Note that the eyes of this entity were uncovered by simply equalising shadow and light. No structure is added to any of the images presented.

It is highly probable that too many people who visited this site have simply skimmed over the 'rocks and boulders' without realising that all of these things are actually the remains of a previous civilisation that once existed here.

It would appear as though everything was destroyed in an instant, reducing once-living organisms, creatures and structures to dust, rubble and rock fragments.

Dark Associates

To find the image of the old wizard-looking being above, the raw picture must be rotated counter-clockwise by 90°

Surprisingly, when examining some of the detail, the preservation of a few of these features is quite remarkable. Could it be due to the fact that there was no bacterial interaction assisting in the decomposition of organic matter that was obliterated in some cateclysmic event, ending all life on Mars?

It seems feasible then that much of the preservation evident is as a result of the lack of bacteria and severe weather erosion (apart from wind in the form of huge dust storms, dust-devils and whirlwinds)

The image below, found in a slightly different location, shows an entity buried up to the chin with what appears to be a clearly manufactured twisted rod with a clover-leaf shamrock design jutting out just to the left of a hooded head. Could this represent a sceptre or perhaps a wand? Many other strange things can be seen scattered in this locality.

Winged Hooded Being with Sceptre / Wand

Raw Image	Enhanced image

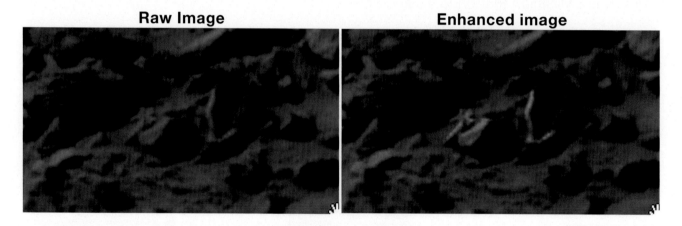

Further examination of the terrain reveals many other things and certainly not what may have been expected in terms of 'Martians' or our idea of alternative life forms living on other planets.

This, however is far more profound and undoubtedly reveals a chapter in our history that has been hidden and will remain hidden to the masses for as long as humanity wishes to remain in the darkness, brought about by the subtle manipulation of our consciousness through mainstream media, science, religion and politics – including NASA.

Winged Hooded Being with Sceptre or Wand

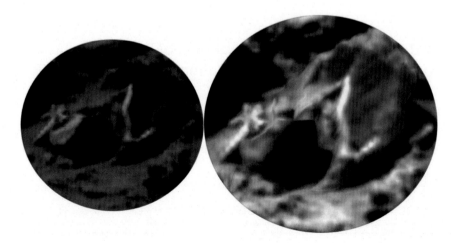

Usually, a **sceptre** is a symbolic ornamental staff held in the hand by a ruling monarch as an item of royal or imperial insignia. It can also sometimes be used for indicating a sense of divinity.

On the other hand, ***magic wands*** commonly feature in works of fantasy-fiction as spell-casting tools. In traditional tales of fantasy, wands fill basically the same role as a **wizard staff**.

Canine-like Creature

Raw Image **Enhanced image**

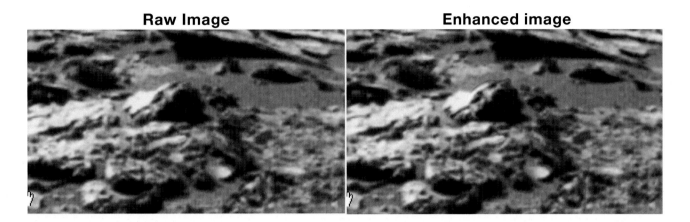

The above picture depicts a head of a canine-like creature. Careful examination reveals many other things also scattered among the debris. The easiest way to identify these is purely through looking for symmetry in the features observed.

It is interesting to note that a large number of creatures resembling dogs were discovered on Mars, not all of which have been included in this book. How could 'man's best friend' be found on another planet, a distance of approximately 225 million kilometres (140 million miles) away, when supposedly these animals had evolved on Earth?

Sadly, the scientific viewpoint is dominant in the world we currently live in and we are only allowed to believe what science determines on our behalf. Yet in spite of this, at the coming of this age (the Age of Aquarius), the secrets of the universe will be revealed to every inquisitive mind.

Zoomed in view showing the head of a Canine-like creature

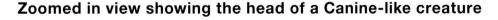

The biggest problem with the modern human psyche is that it cannot differentiate fact from fiction without the help from some 'authority' or 'expert' to confirm what is up or down. We need look no further than the conditioning of the human mind, through suppression of free will, enforced through the legal system, education, and of course religion.

Jackal-like animal and Artefacts

Raw Image   **Enhanced image**

The screenshot above shows a jackal-like animal with other artefacts / creatures that can be seen lying in a pile. A few have been selected and enhanced to bring out the detail.

How much more proof do we need to show that Mars was once inhabited and suffered a disaster wiping everything out in an instant? Why was the official history-making discovery by NASA suddenly changed to another story? Who controls this knowledge of what existed in the past on the Planet of War??

This will serve as a rude awakening in the face of the intellectual elite, including our religious leaders who enjoy deciding what is acceptable for the rest of humanity to think or believe.

Sadly, such is easily accomplished through the various institutions that are supported by the mainstream academia through publishing all the various information necessary to keep us focused in a certain direction. Any attempt to sidestep this path is immediately critised or debunked by them.

Jackal and Artefacts

Artefacts

Strange Face with Headdress

Small Naked Woman

Bell or Hat

Little Face

Exhausted Mastiff

Raw Image

Enhanced image

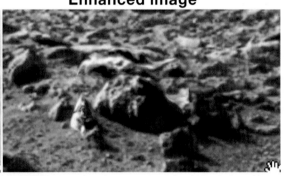

Among the other artefacts and creatures scattered around this area, the enhanced image shows features of a Mastiff-type dog with its tongue hanging out.

Refer to the original image (raw image) above on the left to compare features showing that the structure of the image was not altered. Only colour and contrast were equalised to give more clarity to the detail.

The images below confirm exactly what can be seen:

Toy?

 Ghostly Face

Mastiff-like Dog

 Small horned being

Pair of Feet

Raw Image   **Colour enhanced**

The face rotated by approximately 45° reveals the stark reality of the truth before us and cannot be denied. History, religion and science, specifically cosmology and anthropology will clearly need to be reviewed in order to truly understand our origins. What will the future bring without knowing the truth – and how will it change once we do?

How all of life got to be destroyed on Mars may remain a mystery still to be revealed. However, my personal belief is that this was an intervention of sorts to protect the future interests of humanity on Earth. Is it possible that remaining oblivious of our true history may bring about our utter and complete destruction because we choose to remain, or are purposely kept ignorant?

Phantom in Rock

Raw Image   **Enhanced image**

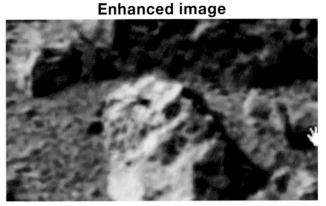

Above is an image of a strange phantom-like face of a being that would cause many of us to shriek in horror if encountered. In spite of the mythical appearance of some of these beings found here, Mars appears to have been a home supporting many other species. Could science and religion provide some answers without any contention…?

Humanity cannot continue to have the likes of 'authorities' dictating and making the rules in terms of what they do or do not want us to know.

Every human has the right to this knowledge in order to consider their own choices and outcomes based on the evidence presented. This is not a kindergarten party after all, this is serious information that may carry eternal consequences.

Below, an enhanced view of the image tilted upright showing the strange eerie, phantom-like face of a horned being.

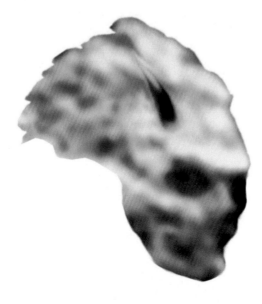

It appears that this was the reality of the Red Planet: it supported abundant life, including a species most humans would consider demonic. These entities appear to have been organic (not spirits or ghosts), and of a flesh determined by the DNA molecule. Procreation among these entities was possibly exactly the same as how all humans proliferate on Earth today.

These beings could have populated Mars just like humans populate our own planet. Could these entities have been the creatures, referred to as the 'fallen angels', spoken of in the Bible (Revelation 12:7–10, NIV), which fled and came to Earth following a war that took place 'in Heaven'?

Devilish Martian

Raw Image   **Enhanced image**

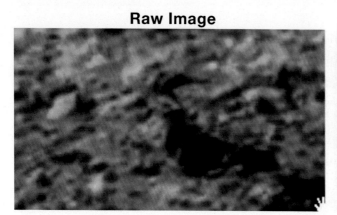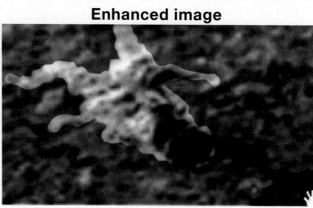

Above can be seen what appears to be a face of a devil with a half-closed eye. It is actually a dead devil, its face embedded on the surface of Mars. What appears to have been a catastrophic event wiped everything out on this planet in an instant, leaving us with only images such as these as evidence of this occurence.

This could have happened a mere few thousand years ago, judging by the preservation of some of these life forms as depicted in these images.

Other filters employed uncover more detail as seen below. This is *not* a drawing.

Devilish Martian

Do we still have any volunteers who would like to spend the rest of their lives on Mars, rejuvenating the planet and restoring its obliterated past? If this image is anything to go by, it might be a far better idea to rather remain on Earth and help restore our own planet.

Thankfully, all these entities are dead. Everything appears to have been laid waste by what could have been a meteorite, asteroid or possibly even some nuclear holocaust. This might be good news for the sake of those who wish to travel to Mars in the future; however, I personally would not be keen to be among those brave enough to venture into this frightfully hellish world strewn with the remains of demons, devils and dark entities that once inhabited the Red Planet.

The Peering Demon

Raw image

The highlighted section showing the face of a demon.

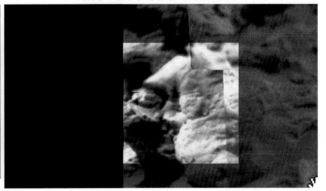

The view above was found by panning to the extreme left where the camera angle is limited. This is indicated by the black portion on the photo towards the bottom left. Mars was without any doubt inhabited by beings and entities that today most humans would consider mythical or folklore.

Here follows a description of how the hidden details are uncovered by using a simple technique.

The raw image clearly shows how this image has possibly been 'doctored' or overlaid with other layers to hide the detail. The enhanced image reveals a demon-like being peering from behind some rubble. The eye on the left is clearly visible with its pupil seemingly staring directly at the camera.

Below, the same picture seen on the left was adjusted to reduce the colour saturation in an attempt to increase the visibility of the other eye. The picture on the right is an inverted view of the left-hand picture, bringing out the features of the other eye that can now be seen, including the detail of the pupil and tip of the tail.

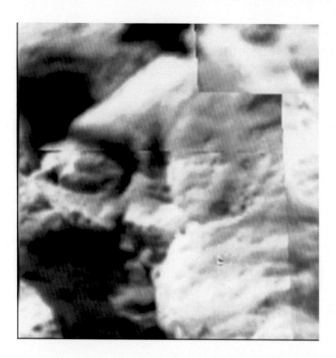 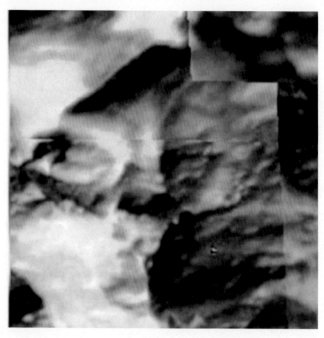

These are simple techniques that anyone can accomplish by utilising suitable software that is able to perform these tasks. This is not 'photoshopping', since no extra structure is added or reduced to enhance the hidden features.

By enhancing (equalising) the visible features seen above, this brings out a rather crude-looking face of what can be considered a demon. This is clearly seen in the picture on the right.

There is no doubt about it: Mars was once a habitat of beings and creatures that most people would not feel comfortable having as their next-door neighbours.

This reality needs to be seriously contemplated in terms of accepting the fact that demons and devils previously inhabited Mars.

What manner of species were these beings? These creatures appear to have been composed of flesh and blood and possibly procreated just like all other living organisms.

What, then, adulterated the DNA Code to produce such creatures that most humans would consider to be the fruit of darkness ... and what purpose did they serve??

Note the tip of the tail .

These are serious aspects to be considered; to understand the universe we live in and its composition.

On the Mars interactive site, the following view was located by zooming in to the exact position where the hand-icon is seen. When zoomed in, a head of a Satyr-like being is seen lying on its side.

Once located, the image was secured by using the printscreen option. The picture is tilted upright to obtain the same view as the one seen below. This again is another example of the reality of what actually existed on Mars, 'The Planet of War', as it was referred to by the ancients.

Raw image　　　　　　　　　　　**Highlighted image**

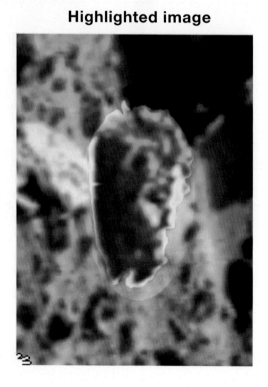

Here, a being resembling a Satyr with Pan-like characteristics clearly shows how everything we were ever taught concerning mythology, fables and folklore appears to have had its place in history.

The Martian Satyr

So what say we now, Mr Richard Dawkins...?

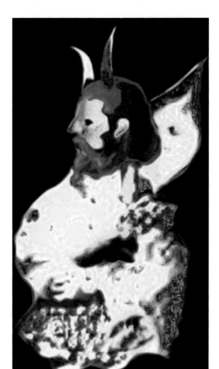

I trust that the reader will now understand that not everything we hear or see from the mainstream media is necessarily true. Here before us is EVIDENCE that life had once flourished on Mars, yet NASA is still totally silent about this.

This discovery on the Red Planet was the start of an epic journey that changed my own consciousness and understanding of the universe, including our miniscule existence in the solar system.

A few days after discovering these things on Mars, I happened to be on Google Earth, scanning the Antarctic terrain, when I came across something that changed my mind about the universe forever.

Having discovered two unrelated aspects in such quick succession, this jolted me to the realisation of how my consciousness had been altered, suddenly having an insight of things I had never previously considered possible.

I could also not help thinking about the Mayan Calender and whether I was a victim of its prediction... Could this indeed be the Apocalypse, an exposé concerning the future of humanity? I decided I needed to dig further...

CHAPTER 2

ALTERNATIVE HISTORY

ABORIGINAL MYTHOLOGY SAYS:

'We were cast out into darkness and were much afraid, so the *Gods* gave us a sun to warm us and a moon to see at night.'

This chapter is dedicated to a specific time in our history that may give us a clue to what we are looking at. Now, consider tutelage as a fundamental diversion, particularly where scholars are exposed to mainstream ideologies, mostly taught in schools, colleges and universities.

Most of history as we know it has its roots stemming from ancient stories and myths carried down from generation to generation, first by word of mouth and later the written word. Initially, the written word became an authoritative and prevailing medium used to introduce freshmen to a formidable communication reflecting actual events, transactions etc., all recorded accurately to ensure its legitimacy. Today however, interpretation and classification of ancient text is used to categorise myth, fable, legend and actuality. So, what is truth and what is folklore, and *who* decides which is which?

Part of the Epic of Gilgamesh

Image Credits: www.revolvy.com

The written word is said to have originated from Ancient Mesopotamia, being inscribed on clay tablets by the first known civilization called the Sumerians using a writing system known as Cuneiform in which wedge-shaped impressions were made in soft clay.

Today, thanks to the internet, accessibility to information is available at the flick of a switch, affording us the opportunity to research phases of our history and to examine at our leisure the deciphered accounts of tales inscribed upon these tablets, mostly considered mythical in nature such as the popular *Epic of Gilgamesh*, also known as 'The Flood Tablet' excavated from the Library of Ashurbanipal, Nineveh (Iraq), housed in the British Museum.

This tablet was recovered by Austen Henry Layard in 1849 in fragmentary form and published in 1876 by George Smith of the British Museum. His translation of this sacred Babylonian epic pieced together from broken clay tablets caused international headlines with his translation of a flood text which paralleled the Biblical tale.

Lesser-known stories (meaning little exposure in mainstream education institutions) include the **Enûma Eliš** (also spelled 'Enuma Elish'), which scholars refer to as the *Babylonian Creation Myths*. The Enuma Elish appears to represent a creation myth that is far more detailed than the brief Biblical account found in the Book of Genesis. Nevertheless, for the past one hundred years, the Enuma Elish has been dismissed by the so-called 'experts' as mere mythology – an imaginative tale of a great cosmic battle of good and evil – a story of battles between one 'god' and another, the hero of which was Marduk, the chief deity of the Babylonians.

The tablet below shows the Sumerian version of our Solar System as they knew it approximately 4500 BCE.

Source: https://i.pinimg.com

In this illustration, the Sumerians depict the solar system, showing the sun and the planets, including the moon and a tenth planet they called Nibiru ('Planet of the Crossing'), also referred to as Marduk by the Babylonians. Keep in mind that these ancient astronomers supposedly never had telescopes.

The Enuma Elish has about a thousand lines and is recorded on seven clay tablets, each holding between 115 and 170 lines of Sumero-Akkadian Cuneiform script. However, most of Tablet 5 has never been recovered, but aside from this gap, the text is almost complete.

This epic is one of the most important sources for appreciating the Babylonian worldview focussed on the supremacy of Marduk and the creation of humankind for the service of the gods. Its primary original purpose was not an exposition of theology, but rather the elevation of Marduk, the chief god of Babylon above the other Mesopotamian gods.

The Babylonian epic lists the 'gods' that were begotten by AP.SU (the sun), with descriptions that match the planets of the solar system in amazing detail. Within this cosmology, the 'gods' anthropomorphised the planets, i.e. having consciousness, will and the ability to act.

An example taken from the texts depicting a great celestial battle between Tiamat (the Earth) and Nibiru (a rogue planet) reads as follows:

'Then, "in the heart of the deep" a new and more powerful god, called Nibiru, was created:

Perfect were his members beyond comprehension... unsuited for understanding, difficult to perceive. Four were his eyes, four were his ears; when he moved his lips, fire blazed forth... He was the loftiest of the gods, surpassing was his stature; his members were enormous, he was exceedingly tall.

Tiamat and Nibiru, the wisest of the gods, advanced against one another; they pressed on to single combat, they approached for battle armed with a "blazing flame" and having acquired various "winds" (or satellites), Nibiru "towards the raging Tiamat set his face".

Nibiru spread out his net to enfold her; the Evil Wind, the rearmost, he unleashed at her face. As Tiamat opened her mouth, to devour him, he drove in the Evil Wind so that she closed not her lips. The fierce storm winds then charged her belly; her body became distended; her mouth had opened wide. He shot there through an arrow, it tore her belly; it cut through her insides, tore into her womb.

Having thus subdued her, her life-breath he extinguished. After he had slain Tiamat, the leader, her band was shattered, her host broken up. The gods, her helpers who marched at her side, trembling with fear, turned their backs about to save and preserve their lives. Thrown into the net, they found themselves ensnared... The whole band of demons that had marched on her side he cast into fetters, their hands he bound... Tightly encircled (gravity?), they could not escape.'

The *'planet'* Tiamat was thus 'extinguished', but the act of creation was not yet finished. Nibiru became caught in the orbit of the sun, forever to return to the place of the celestial battle with Tiamat. On the first encounter, Nibiru's satellite winds (moons) had smashed into Tiamat, but one orbital period later, Nibiru itself 'returned to Tiamat, whom he had subdued' and the two 'planets' did collide.

Here, among other things is what some scholars believe to be a cosmogony, a description of the formation of our solar system and, in particular, how Earth found its place and current orbit around the sun. Described in the text above is a great celestial battle between Tiamat, which became known as Earth after the battle, and Nibiru (Planet of the Crossing as it was aptly named by the ancient Sumerians), which intercepted Tiamat's orbital path around the sun resulting in a collision between the two, some 4.2 billion years ago. The Babylonians, following the Sumerians, later renamed the planet in honour of their supreme god, calling it Marduk.

Nibiru vanquishes Tiamat by splitting her in two halves exposing her golden veins, one portion eventually becoming the Earth, and then obliterating her head into fragments, ultimately to become known as the Hammered Bracelet (the Asteroid Belt). Nibiru was given the right to set his own destiny – which, according to the writings and interpretation of the clay tablets by Zachariah Sitchin, is referred to as Planet X, moving around our sun in an elliptical orbit every 3,600 years.

The point that is been emphasised here is this: the earliest texts refer to the sun and the planets as 'gods', i.e. living entities which rule the solar system. This should be noted, especially considering the other aspects that will later be revealed in this book.

The Mayan Calendar and the Age of Aquarius (a brief overview)

Image Credits: https://cdn1.vectorstock.com

For those who are familiar with the Mayan Calendar and the so-called Age of Aquarius, here is something to consider. It is understood, according to experts, that the Mayan Calendar ended on the 21st December 2012; the supposed prophesy that the world would end on that date was well publicised by various esoteric groups, some of whom had their own interpretation that this would have been the end of the world.

And yet, we are all still here and nothing even remotely like that happened. It was, however, an apocalyptic event, as we will see later.

Others had a different interpretation regarding this 'Count of Time' though and declared that this was simply a shift in consciousness brought about by the emergence of the purported Age of Aquarius beginning the 22nd December 2012.

Some scholars of the Mayan Calendar say that we have already completed one of the most important Mayan calendars of all, known as the ***Tun*** on 28th October 2011, which measured the flow of consciousness in nine distinct levels as related to time.

The **Tun** calendar is 16.4 billion years long and goes all the way back to when the universe was still forming. This is not coincidental, as many debunkers would want us to believe. The Mayan's were totally obsessed with time (our 4th dimension) and had calculated complex cycles of which the biggest would end on 21st December 2012.

The following day, 22nd December, marked the beginning of the Age of Aquarius, the revealing of the 5th dimension – i.e. the Apocalypse.

In astronomical terms, the end of 2012 also marked the end of a cycle known as the 'Precession of the Equinoxes'. Such a motion consists of a cyclic wobbling in the orientation of Earth's axis of rotation.

Currently, this annual motion is about 50.3 seconds of arc per year or 1 degree every 71.6 years. The process is slow but cumulative and takes 25,772 years for a full precession to occur. This has historically been referred to as the Precession of the Equinoxes (see below).

The discovery of this cycle is attributed to the Greek astronomer Hipparchus, who determined that the positions of the stars did not match up with the Babylonian measurements that he was accessing.

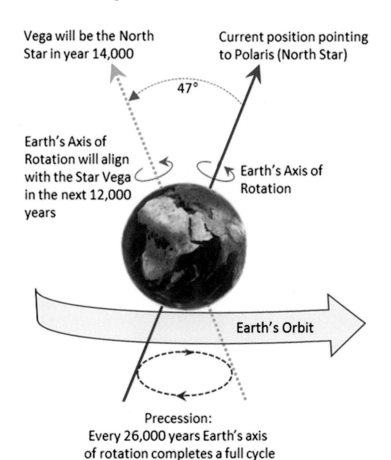

Vega will be the North Star in year 14,000

Current position pointing to Polaris (North Star)

47°

Earth's Axis of Rotation will align with the Star Vega in the next 12,000 years

Earth's Axis of Rotation

Earth's Orbit

Precession:
Every 26,000 years Earth's axis of rotation completes a full cycle

According to the Chaldean chronicles, the stars had shifted in a rather organised way, signifying that it was not the stars themselves that had moved but the frame of reference – in other words, the Earth itself.

The key word here is 'cycles' and could be extremely significant because, as one cycle ends, another begins. The beginning and ending of cycles could be linked to an **event** mentioned in the later chapters of this book.

Interestingly, there has been much speculation that the Mesoamerican Long Count Calendar is somehow calibrated against the precession, but this view is not held by professional scholars of Mayan civilization.

Here lies all secrets known only by the initiated, including the characteristic of magic and witchcraft that now will become unveiled in order that the prudent can be made aware of its concealed nature. This book will testify to this fact.

I am by no means propagating the use of these dark practices, which are forbidden, but rather to bring about awareness of this reality, especially since this book presents some of the most startling revelations in this respect.

The reader will perceive how strongholds over humanity are formed because all matter consists of the same 'dust' that creates our flesh and consciousness, including the atoms that make up our DNA. We inherit from our forefathers elements that determine our physical appearance, including memories that they experienced, which ultimately impact our current level of consciousness and our behaviour and the future to come.

Furthermore, according to astronomers, the entire solar system completes one orbit (known as a Cosmic Year) in 225 to 250 million years around the Milky Way while orbiting the Galactic Centre. During this journey, the solar system oscillates and passes through the **central equatorial plane** of our galaxy approximately every 13,000 years. According to experts on the subject, the Earth has now entered the equatorial plane, which is charged with photons (Photon Belt) that some believe will begin to alter the DNA of all living things, bringing about a significant shift in consciousness and usher in the so-called Golden Age.

So what then are photons? All light is comprised of photons, which are tiny particles far too small to see individually. According to science, the earliest photons probably appeared about 15 billion years ago during the so-called 'Big Bang' (which remains just a theory, and a contentious one at that, as this book will show). Unlike electrons and quarks, photons have no mass so they can travel at the speed of light (about 297,600 kilometres per second).

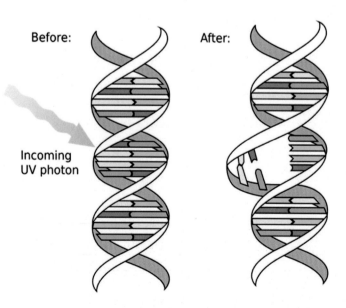

Before: After:

Incoming
UV photon

Image Source: https://upload.wikimedia.org

Photons behave in some ways like particles, and in other ways like waves. It is not just visible sunlight that is made of photons, but many other waves including radio waves, television transmits, x-rays and the ultraviolet rays that give us sunburn.

The difference between visible light and the other types of waves depends on the size of the wave – i.e. the wavelength. X-rays and ultraviolet rays are very short waves that can damage our skin and our DNA. Visible light like sunlight is comprised of medium-length waves, while radio and television waves are very long waves. All of these rays are made of photons.

The diagram above portrays the potentially harmful effects of UV on the DNA structure. If a photon moves in short wavelengths, such as x-rays or ultraviolet rays, then that photon has more energy and hence more harmful effects.

If the wavelength is longer though, then that photon will have less energy – which is why it is less dangerous to stand in the way of microwaves, cell phones, or radio or television broadcasts.

When photons bump into other atoms, some of their energy can cause the electrons in those atoms to move faster than they were before, which is what we call heat and explains why we get hot while sitting in the sun. (Microwaves work differently: they are long waves that hit water molecules just right to make them flip around, and it's the friction from the water molecules flipping that heats your food.)

However, this is not intended to be a science lesson, but rather to create an awareness of the possibility of a Photon Belt extending across the equatorial plain of the Milky Way galaxy. A superwave, or a plain where only pure photon light exists, is a place in time where possibly all atoms of which physical matter is comprised become inert, all motion ceases (Newton's Third Law of Thermodynamics) and all matter crystallises in an instant.

Could this observation of former living biological matter, now turned to rock as observed in this book, possibly indicate how the earth, solar system and the universe were subjected to some kind of cataclysmic event, taking place over various cycles of time? While the notion of the Photon Belt may merely be a part of New Age philosophy, some aspects of the story can be analysed scientifically.

Scientists believe that should a 'Photon Belt' be physically possible, it would require the gravitational pull of a black hole, with light rays being bent around the black hole near the event horizon, forming a 'photon sphere'. [Note: The very central point of our galaxy is a black hole, so the concept is plausible.]

In June of 1979, Dr Paul LaViolette deciphered an ancient constellation message describing the past arrival of a cosmic ray volley from our galaxy's core and of its subsequent cataclysmic effect on the Earth. The following month he wrote this up as a short paper on this 'superwave' concept.

In 1983, after four years of Ph.D. research, he published his investigation of these galactic superwaves and their connection with cyclic global cataclysms. In this dissertation and his subsequent papers and books (*Beyond the Big Bang* and *Earth under Fire*), LaViolette appears to have justified every statement he has made.

LaViolette's scientific papers describe evidence of galactic cosmic ray superwaves emanating from the Galactic Centre, each outwardly moving superwave shell producing a ring of electromagnetic radiation concentric with this centre and lying along the galactic plane, accompanying the superwave as it travels outward. This radiation zone could be termed a 'Photon Band' or 'Photon Belt'.

LaViolette has shown that radiation coming from the nearest of these superwave radiation rings, at its closest point to us, would appear to originate from a region lying about 7,000 light years away in the Taurus constellation region (~6500 light years further away than the Pleiades). In the opposite direction, toward the Galactic Centre (Scorpius constellation region), it would lie furthest from us, about 30,000 light years away.

Although, from this perspective, the Photon Band concept lacks supporting evidence, perhaps the observations presented in this book may provide a clue and create an interest in trying to learn more about the superwave concept.

Leading experts of both quantum science and the Mayan Calendar are saying that our DNA will be 'upgraded' (coded with intelligence) from the centre of our galaxy – the 'Great Central Sun'. According to those who have studied this emerging dogma, the following is a typical example of how certain 'experts' on the subject have predicted its inevitable impact on human consciousness:

'Robert Stanley reported on the discovery of the Photon Band by satellites in 1991 and commented: "These excesses of photons are being emitted from the centre of our Galaxy... Our solar system enters this area of our Galaxy every 11,000 years and then passes through for 2,000 years while completing its 26,000-year galactic orbit"' – from **The Photon Belt**, Part 1.

The Mayan Calendar is the only calendar known to be based on Galactic Cycles. The Mayans claimed they created this calendar to monitor the light coming from the centre of the Galaxy as affects our DNA. We now know through the work of Dr. Fritz Albert Popp, that DNA not only absorbs light but also emits light. DNA also appears to be the bridge between our physical and etheric bodies. Modern science now realises that our DNA directly reflects our consciousness, making it possible for us to wilfully change our DNA' – from **The Photon Belt**, Part 2.

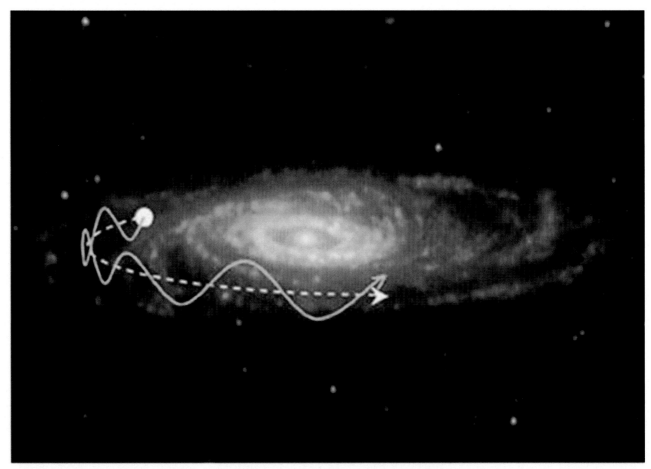

Image Source: https://i.stack.imgur.com/Dv1li

Note that since the end of 2012, our entire solar system supposedly started moving through the equatorial plane of our Milky Way galaxy, an area charged with photons, and this will intensify as we approach the precise centre.

During this traverse, our solar system will oscillate in and out of the Photon Belt approximately every 13,000 years. The Earth is set to remain within the narrow band of the Photon Belt situated across the equatorial plane (dotted line) for about 2,200 years.

Above is an artist's impression depicting the orbital path of the entire solar system travelling around the Milky Way galaxy.

Some of you may remember, back in 1967, the Motown group The Fifth Dimension and their hit song called 'Aquarius / Let the Sun Shine In'. This song became the collective cry for the sixties hippy generation, pointing to the emergence of the Age of Aquarius.

The words to this song also present some foretelling archetypes for the times we find ourselves in:

Harmony and understanding
Sympathy and trust abounding
No more falsehoods or derisions
Golden living dreams of visions
Mystic crystal revelation
and the mind's true liberation
Aquarius! Aquarius!

Was there ever a more perfect group to sing it than one named The Fifth Dimension... or is this merely a fluke?

According to followers of this esoteric knowledge, the Age of Aquarius will herald the time when humans will transition from a reliance on the conscious mind to the heart instead. It is the transformation out of the ego or 'me' consciousness to a more highly evolved 'our' consciousness or oneness, whereby humans will become much more concerned with the experience of life rather than the possession of material things.

Furthermore, the 5th Dimension exists within the human psyche as a projection from the DNA itself. When I first heard about the Mayan Calendar's 'End of Time' and the subsequent 'Shift in Consciousness' that was going to occur after the 21st December 2012, I wondered incredulously about this event and whether it would really unfold... as predicted by the Mayans in their renowned calendar. The next chapter presents some of my findings and conclusions in this regard.

CHAPTER 3

THE APOCALYPSE

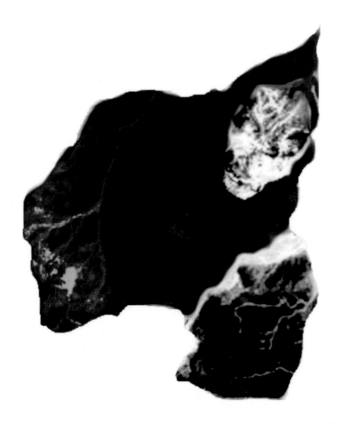

This chapter is not a 'conspiracy theory'. It is the revealing of things long hidden from the ordinary man in the street. Despite the Earth's current disposition due to erosion and seismic activity, the true hidden character of the planet can be uncovered by using simple filtering techniques. It is Apocalyptic in nature revealing past events and cycles and the possible fate of humankind, presented with unquestionable evidence that you the reader may also validate by utilising similar methodology. The choice is yours.

After the discovery of what existed on Mars, this led to seeking further evidence on Earth (utilising Google Earth), which unwittingly resulted in making a far bigger discovery. Indeed, this discovery, by comparison, made the Mars enigma seem somewhat insignificant.

Consequently, since this discovery on Mars was the beginning of a journey that led to uncovering a greater mystery, Chapter 1 is dedicated to the Red Planet and provides a possible explanation concerning the resignation of Pope Benedict XVI of the Catholic Church, in February 2013.

This and the following chapters contain the most challenging information ever to be considered, so much so that it is going to change the rules of our perception, particularly our understanding and view of the Earth, the solar system and ultimately the universe. When all the evidence in this book is taken seriously, a significant shift in consciousness will occur and a new genre will develop, changing our philosophy about the universe and our place within it.

It is certain, however, that there will be an attempt to suppress and debunk the information contained herein. Regardless of this expected knee-jerk reaction, the shocking truth *will* prevail, and there will undoubtedly be a shift in consciousness and the hidden history of the universe (at least a portion of it) will be uncovered, setting a new realisation concerning our past. Nothing like this has ever been contemplated by either science or religion. Ironically, both appear to have missed the reality of the creation.

As you read this book and view the pictures that are hereby presented, it becomes clear how easily the attention of humanity has been distracted from possessing this knowledge. This deviation is accomplished through schools and universities (education is a fundamental diversion), with further distractions through the mainstream media.

Furthermore, all of us focus on the need to survive in a world that demands our attention to fulfil Maslow's Hierarchical Needs, i.e. the activities we perform to ensure that our basic needs are met, including water, food, shelter, security etc., all of which is governed within the 4th dimension that we call time.

Politics and religion play the biggest role in keeping the masses ensnared in a world riddled with conflict, hatred, deception and lies – and alas, we are content to accept the status quo as long as we can survive and provide for our basic needs.

Most of us unquestioningly agree or consent to whatever the authorities and experts convey, typically something that they themselves have been taught. On the other hand, if it is not vetted by an expert, it will unlikely be accepted as fact, since anything that goes against or challenges the status quo is debunked and ridiculed, mostly by the establishment who have a vested interest to protect their own veracity.

Humanity is obliged to accept the existing state of affairs controlled through and by institutions dealing with educating the masses. But the time has come for big and very necessary changes…

This knowledge will pose the biggest challenge to science, forcing the foundation to rethink theories concerning the origins of the universe our solar system and evolution. The evidence presented will also place all religious institutions under the spotlight and simultaneously challenge the atheist to review both dogma and convictions pertaining to faith or disbelief. Every doctrine will be tested, and many of these will fade into insignificance. Politics will change, a new philosophy will emerge, and this knowledge will profoundly alter the present consciousness of humankind.

There will be huge debates and various analyses and interpretations of these findings; however, anyone who views and confirms the validity of the evidence presented will indeed experience a profound change in consciousness; previous views and perceptions of our existence will be altered, while the solar system and the entire universe will be seen in a different light. Indeed, it will place the reader within the scope of realising another dimension of space and time never considered before.

In the following chapters, the evidence presented consists of screenshots obtained from Google Earth, NASA's New Horizons' probe and the Hubble Space Telescope. It must be noted, however, that the detail of the images presented in this book have been enhanced and do not appear as such directly from any of the raw images available from these sources. Detail is uncovered by applying various filters and techniques thereby revealing the evidence.

Astronauts in orbit around the Earth may never have noticed these features, perhaps due to cloud cover, and if they had, it was never brought to the attention of the general populace. The features reveal a shocking fact: our entire planet is the product of organic matter that once existed in gigantic proportions consisting of various life forms, some of which are still recognisable on Earth today, such as birds, snakes, rabbits and even insects.

The coordinates of the various sites that are displayed within these pages are visible on all the Google Earth screenshots, so the reader is therefore encouraged to investigate the reliability and validity of these findings; and, through so doing, concluding that these aspects can in fact be noted and enhanced when similar methods are utilised to reveal the underlying detail. It is important to note though that in certain cases, the emphasised features that appear after 'treatment' may need to be subjected to considerable effort and time to bring out the finest of detail; however, in most cases, the process is simple.

I want to emphasise I am NOT an artist, and the images presented have **not** been drawn. Anybody can uncover these images by carrying out a similar modus operandi. No attempt was made to alter the structure or shape of any of these images. In certain cases, different colours / inverting colour and shade etc. are employed to identify / differentiate specific features apart from others, but never changing the actual shape or adding extra structure that would compromise the validity of the image. I apply a simple technique that anyone can utilise, uncovering the detail revealing these incredible images. The raw images are mostly hidden; however, close observation will reveal the disturbing reality of what makes up the substance of the Earth and all the other rocky planets as well. In certain cases, it may not be necessary to apply any of these techniques as many details are clearly visible, even without the need for further adjustments to light, colour or shade.

In all cases, the reader is encouraged to remain objective and with an open mind. Note also that in terms of probability and statistical significance, most hypotheses are tested against observations. Randomly sampling any given area on Earth will prove the significance of these findings. In fact, these aspects appear over the entire solar system and universe and are not simply isolated shapes *resembling* a recognisable attribute, but rather actual features that once belonged to living entities from the past. The frequent discovery of these things, including the detail and clarity obtained, is enough to convince any person with an average intelligence that this is fact.

I encourage anyone who dismisses what they see here as simply an illusion to consider engaging the same or a similar technique as I have employed to demonstrate this reality. While our present state of consciousness keeps us entangled in the 3rd and 4th Dimension, the hidden mysteries of the universe are nevertheless about to be unlocked, revealing its secrets that will ultimately confound us all.

For the open minded, it will be an amazing journey of discovery unravelling a profound mystery where our past and our future are nested within a much greater and grander scheme; one that is far more profound than what any human has likely ever imagined, particularly with regard to the creation of our universe. With this in mind, the images presented capture robust evidence of a **fractal**, incorporating dimensions of space-time and repeating patterns that existed in the past, and are again to be repeated in the future. As we look further back into time, these patterns can be traced, showing life forms on a magnitude that would appear impossible. Our current science and understanding of the cosmos are due for review.

Here, myth, fairy tale and folklore are fact and no longer mere fantasies of the imagination…

Welcome to the 5th Dimension!

CHAPTER 4

GIANTS ON EARTH

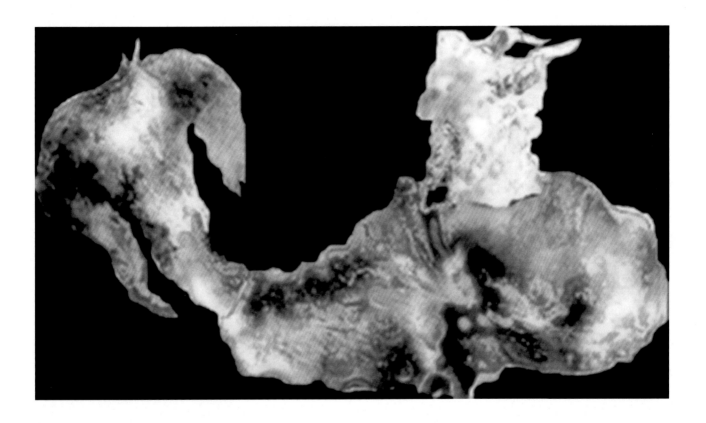

ECCLESIASTES 1:9 (NIV)

What has been will be again, what has been done will be done again; there is nothing new under the Sun.

A DIFFERENT TIME, ANOTHER SPACE ....

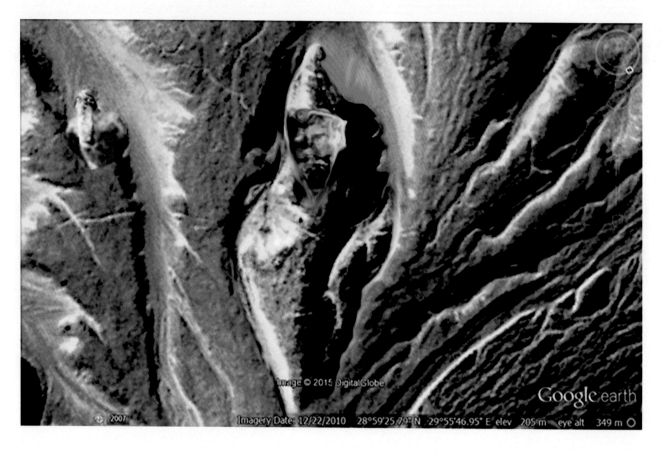

When we speak about giants, most of us may reflect on the story of David and Goliath, the nine-foot giant who was conquered by David in the renowned battle described in the Bible.

However, the giants revealed in this book, by comparison, make Goliath appear more like a microbe (literally!).

These beings, found in Egypt, are a typical example of how the formation of the Earth including its continents, mountain ranges and valleys are a direct result of gigantic biological entities that once existed in a space and time we are yet to understand.

The larger entity (I call Merlin of Egypt) would have stood approximately 135 metres tall. On his left, a smaller entity is noticed with his arms behind his back as if in a pose, showing off his muscles. However, this could also be a possible depiction of submission to the larger entity. Whichever way one may want to interpret this view, the fact remains that these beings were enormous by comparison against any previously-noted giants, whether based on contentious 'conspiracy theories' or otherwise factual data. Tyrannosaurus Rex would appear as a large lizard from Merlin's perspective.

Merlin of Egypt

Before Dinosaurs roamed on the Earth

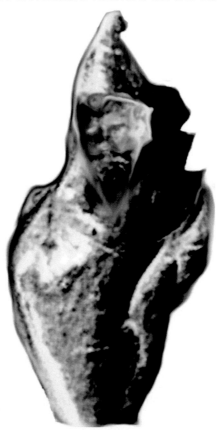

Merlin

Approximately
135 meters tall.

Muscle Man

Approximately
41 meters tall.

How is this possible?

Another striking feature noted is that each aspect observed appears to have been captured in a 'photograph' or snapshot whilst seemingly busy or animate. Caught unaware, they suddenly became inert and eventually over time turned into rock, the very substance out of which we derive our livelihood.

This characteristic appears to be a common feature noted in all the pictures displayed, perhaps explaining an event that is repeated over time, stretching over billions and billions of years. There is also supporting evidence showing that all organic matter is being transformed into smaller sizes over time.

The Earth, as we know it today, could never have supported most of the creatures revealed in this book. The planet would be far too small and the atmosphere totally insufficient. In fact, many of the creatures discovered existed long before the solar system as we know it was ever formed.

What has been exposed shows how the Earth is a conglomeration of the remnants of various life forms that once existed in the past. Many of these stem from different epochs, and the evidence suggests that each dimension had its own specific time intervals, scaled proportionately to facilitate, and accommodate the various-sized creatures. The larger aspects could have existed further back in the past and the smaller ones in more recent times, but they may also have existed **simultaneously** in different dimensions, a possibility we cannot exclude.

What we are looking at is a view of a magnificent and mystical past that once permeated a 'Place in Space', long before the solar system as we know it today was formed.

The evidence presented can be verified by the reader, who can visit the exact locations found on the coordinates (Google Earth) and apply a similar method to expose the underlying reality on the surface of the planet.

This is not a conspiracy theory but evidence that you the reader can verify, thereby shutting any attempt to debunk this information and allowing anyone who is intelligent enough to investigate and validate this reality for themselves.

Furthermore, apart from the published pictures, other investigators will find similar features all over the earth, since every continent, mountain range and valley are a direct result of biological matter that existed in the past on a magnitude that is quite simply baffling.

The possibility of some *event* occurring within a cycle cannot be ruled out. This is evident and noticeable as an 'imprint' frozen in time, where all organic matter, at the very onset that the event occurs, is left in a state of inertness, and eventually returns to the dust that forms the planets and stars that make up the universe.

The picture above is what I have named the 'Antarctic Priest'. Note the bearded face and the hood with a tassel (see pages 78–79 for more details), and, as an example, is compared against the size of Merlin, embedded in the circle marked by the arrow. The Antarctic Priest is enormous by comparison (7.3 kilometres tall).

What becomes clear is the fact that taking the current diameter of the Earth, spanning some 12,742 kilometres, and its breathable atmosphere, reaching an average height of about 12 kilometres above the surface, our planet could never have supported these larger entities at any time during its history as we currently understand it.

Furthermore, there is absolutely no doubt that most of the creatures uncovered could never have coexisted in the same space as we do on Earth today. For instance, the food that the Antarctic Priest may have required as sustenance would not have been suitable in the dimension peculiar to Merlin (due to the size discrepancy) and they therefore could not have shared the same time-space together to survive. The Earth must have been much larger during the days of Merlin, and larger still during the days when the Priest was around… or did these aspects coexist in separate worlds that cannot be explained?

I propose that proportionate aspects existed in their own peculiar time-space. Each time-space is arranged within a fractal, designed in accordance with the Golden Ratio 1.618. The entire fractal is the 5th Dimension – an intelligent design. The observations displayed in this book provide the evidence to support the very nature of the fractal, and its repeating patterns (see page 133) and possible connection to the Akashic Record.

We can perceive, for example, how in certain areas where much of civilization has been influenced, the overarching features that are noted within those regions seem to have a direct influence on the historical character and culture of that specific area being observed. The Sinai Desert, for example, is one of great interest with many colossal features, millions of years old, which astonishingly seem to correlate with some of the Biblical accounts pertaining to more recent events. Specifically noted is the Exodus out of Egypt and the Ten Commandments given to Moses by God, which occurred circa 1446 BC.

Indeed, this is but a sample of many other aspects presented in the book showing space-time dimensions peculiar to a fractal, each event ending as if caught in a candid photograph in a split moment of time.

Science will need to reengineer previous theories concerning the Big Bang and the creation of the universe, following which our current philosophies and notions regarding our cosmology and anthropology will take on a new, and far fuller meaning. Consider the current theory that all birds evolved from the dinosaurs. If this is true, then how do we explain the enormous birds (see pages 42–46) discovered in Asia, Arabia and Antarctica, which would make the largest dinosaur ever discovered appear as small as a flea by comparison?

And yet, as we gaze further back into the past, these same birds appear microscopic by comparison to the enormous duck or goose that can be seen on pages 86–87.

Many other birds have also been discovered that were not included in this book. These birds obviously existed billions of years before any microscopic dinosaurs (by comparison) roamed the earth. And not only birds, but numerous other animals as well, of which I have included an elephant and a hare. How will palaeontologists explain this? The evidence presented suggests that there was no such thing as evolution… Monkeys, after all, will remain monkeys and whales will never walk… end of story.

However, both science and the Bible agree that Man is formed 'out of the dust of the Earth', which makes a lot of sense when considering that we are the very substance of the dust of matter that was once part of living organisms that existed in the distant past, now fossilised, providing our minerals, metals, fossil fuels – and even our physical bodies.

We plant seeds in the Earth and eat the fruits of our harvest – the nourishment thereof coming out of the 'dust' within which they are grown. We all live and have our being and owe our existence to the giants that became dust, yet now provide the means of our sustenance derived from their fossilised remains. Every item we own and every commodity we utilise comes out of the Earth, including every single cell of our body.

It may occur to the reader that many of these beings and creatures discovered everywhere on our planet take on the resemblance of mythical life forms that can be found all over the world in storybooks and other fabled writings. This is an interesting observation, since how did any human, specifically those who in the past created either a drawing or a sculpture resembling any of these beings, have any prior knowledge of what these entities would actually look like? Good examples displayed within this book show dragons, demons, angels, wizards etc. looking exactly how a modern-day artist might portray them.

Does this perhaps have something to do with our DNA? After all, we inherit from our parents the very genes we pass on to our children; memories of the past are encapsulated in our DNA, handed down over thousands of generations from our forefathers.

When looking at these life forms, you may be reminded of a past that is embedded deep within the psyche of every human; things we are familiar with, no matter how bizarre they may appear. This is the exposure of our roots pertaining to both good and evil, the Akashic Record in pictures.

For others, this may be the ultimate psychedelic experience and the so-called 'enlightenment' sought by mystics and other followers of esoteric philosophy. However, it was never contemplated on a scale such as this. The final destiny of all flesh is clear – but what about the spirit / consciousness we carry within our DNA?

Humankind is placed unwittingly under a spell of false information being fed through institutions including schools, colleges, universities, churches and the mainstream media. Perhaps the keepers of this knowledge want to maintain a certain order to keep humankind under a type of hypnosis, controlling our destiny. The masses are already conditioned, indoctrinated, confused and incapable to perceive and distinguish deception from truth. Could humanity ever be ready to accept such a profound revelation such as this, revealed at this time in our history?

For the few though who are adept and ready to accept this reality, the evidence in this book will have far-reaching ramifications; revealing the most astonishing aspect that will change our view of the universe forever.

Here, in this book is **evidence** of another dimension we have never been aware of, appearing as a **fractal** of **space-time**. Look and learn and then decide what all of this might mean. I do not have all the answers, and hopefully someone will be brave enough to take the reins and provide some insight as to what may really be happening here. The challenge is for science and religion to unite and provide some answers to the many questions that will be asked once the evidence is analysed and accepted.

All readers who have access to Google Earth are encouraged to use the coordinates given in order to verify these findings and to satisfy themselves that this is as real as it gets.

Bird with Larvae-like Nymph

The picture below is a snapshot taken from Google Earth in a region situated around Kazakhstan, viewing the terrain from a height of approximately 53 kilometres above the earth.

The image clearly shows a bird that was busy eating a larvae-like creature situated at its foot-end, as if the bird is perched on it.

The bird in some ways resembles a kingfisher with a hoopoe-type beak, however, I have not been able to identify the species. This could possibly be confirmed by ornithologists.

The bird seen above is as it would appear on Google Earth, concealed among the mountains and valleys and hidden away from the eyes of humanity purely due to the size that it happens to be.

In all cases presented, the reader can confirm this fact by accessing Google Earth through utilising the given coordinates as seen in the screenshot. Once the spot has been located, the investigator should revert to roughly the same eye altitude, which in this case was approximately 53 kilometres, to provide the same view and resolution. This should be followed by adjusting the compass bearing as seen in the top right-hand corner.

What is interesting is the fact that an entire village with adjacent farmlands all happen to be situated on the bird's head. The inhabitants are obviously totally unaware of what they are living and farming on. The black-and-white-striped larvae-like organism at the foot-end of the bird appears to have been partly devoured, a piece still attached to the bird's beak.

The shape of the birds' left wing and feathers are clearly visible including the detail seen regarding the eye and the beak.

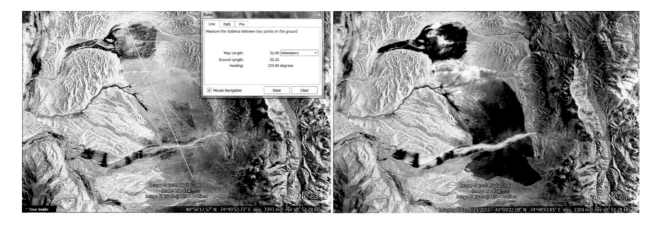

The picture on the top-right portrays a detailed view of the features after subjecting the image to contrast and equalising colour.

This bird is enormous – the yellow line depicted in the picture above left indicates a straight-line distance, measuring approximately 32 kilometres from the top of the head down to the position of the tail. Even flying in an aircraft over this bird at an altitude of 12,000m would not reveal its outline or shape.

Bird with Larvae-like Nymph

Caught in a moment of time before the solar system as we know it today came into being, this bird clearly indicates that a very different solar system must once have existed, somewhat proportionately suitable to support larger creatures such as this.

What caused this enormous bird and larvae-like creature to suddenly freeze and eventually become solid rock, and now form part of the topography of the Earth today?

This is a mystery...

Consider the following: Based on statistics relevant to a certain type of kingfisher, with an average length of approximately 27 centimetres and an airspeed of approximately 57.6 kilometres per hour, scaled up to match the size of the specimen found above, this would mean that it could fly at an airspeed equivalent to 6,826,667 kilometres per hour on our scale! That would mean it could reach the sun, a distance of 149.6 million kilometres from the Earth in 22 hours, travelling at about 0.63% of the speed of light!!

Below is a closer view of the nymph or larvae-like creature the bird was busy eating (perhaps someone can identify this creature?). What stands out is the extraordinary feature of a somewhat three-dimensional quality, also noticeable in many of the other images that are presented in this book. I cannot explain this phenomenon.

However, this picture alone presents sufficient evidence to support the hard facts presented herein. Notice the symmetry of the triangular insectoid head with antennae and bulbous eyes. The twisted flattish body of the creature appears to be partly devoured at the tail end. Who can deny what we are looking at?

Clearly, something is amiss pertaining to our understanding of the cosmos.

The visible feature of this organism measures approximately 28 kilometres across from head to tail in a straight line. This view alone, of the bird and its meal together, emphasises the suddenness of the event, showing that the bird was still engaged in eating its meal and did not attempt to fly away from some obvious catastrophe.

Furthermore, where does this place the theory of evolution? This evidence will immediately question everything that was ever presented regarding this highly-respected theory, upheld by most of our leading academics. Simply because science might not support, or perhaps is restricted when it comes to publishing / verifying information of this nature, this should not discourage the reader to investigate this reality for themselves. Many will find the same and similar things all over our planet.

Findings such as this, together with the other life forms presented, cannot be simply shrugged off as a bad case of pareidolia, a psychological occurrence whereby the mind responds to a certain stimulus by perceiving a familiar pattern where none in fact exists. These features are statistically significant observations. In many cases, multiple features appear supporting group-related activities and other phenomena. Randomly-selected areas at various altitudes chosen anywhere on the earth will show features bearing characteristics that are unique and, in some cases, relate to each other, such as the bird and the larvae-like creature it was busy devouring. Each of these are recognisable, not simply as shapes, but rather for what they actually are.

These aspects are in clear view to anyone willing to undertake an unbiased examination and analysis, using a similar methodology as acquired in this book. Let us not follow like sheep, leaving everything to the 'experts' to take care of what we need to know. If you are only prepared to accept what the experts tell you, then you have lost the plot completely. Do your own research.

The truth will set you free.

Below is another example of a massive bird, in this case found in Arabia. This bird measures about 18.5 kilometres across from head to tail in a straight line and is about half the size of the kingfisher-like bird seen above. The bird viewed from an altitude of about 30 kilometres above the Earth appears to be a bird of prey. The darker feature surrounding the bird could have been blood. I therefore call it Bleeding Bird. This image displays a feature I consider an anomaly, the only feature I have discovered of an animal or being that is captured in an inanimate posture at the moment of a possible 'event'.

Bleeding Bird

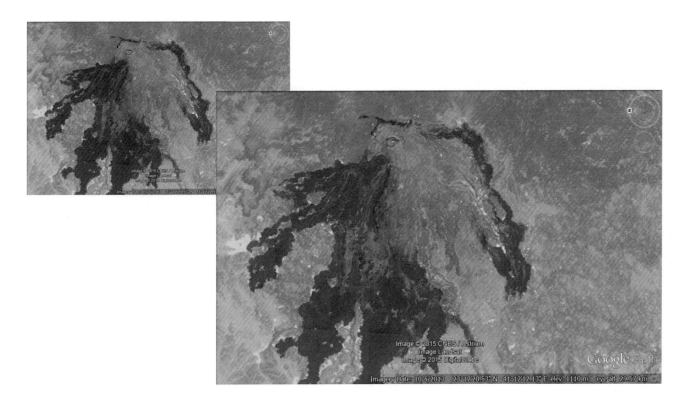

However, the appearance of a broken wing and possible blood suggests that this bird suffered its injuries perhaps just a few moments before the event occurred.

Whichever way we wish to interpret the view of this feature, the fact remains that it was a bird. Notice the head, eyes and beak resembling an owl, falcon or kite.

Sitting Duck

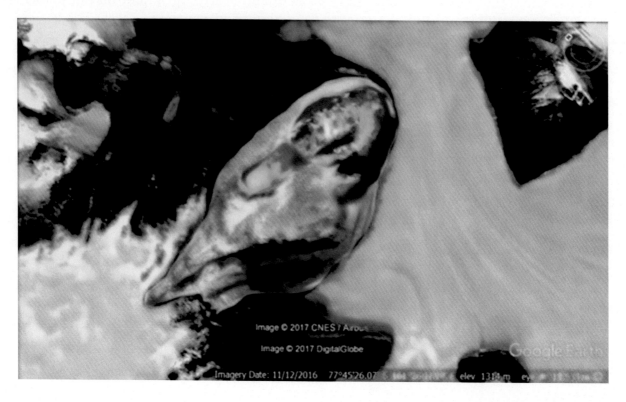

Sitting Duck can be located in Antarctica. What is interesting about this picture is the appearance of flowing water which seems to be splashing at the tail end of the bird. Now, this feature may also provide a clue regarding the 'sudden freeze' theory that I am proposing (see page 27). As mentioned earlier, this is not necessarily a cold freeze, but rather a sudden unexpected event possibly causing all atoms to become instantly inert.

Is it possible that the Earth and the entire Solar System passed through the very central point of the equatorial plain (Photon Belt) of the Milky Way galaxy, as it were, resulting in this event? Could it be the result of the past arrival of a cosmic ray volley from our galaxy's core, a superwave of photons and of its subsequent cataclysmic effect on all matter?

The length of the bird (with head tucked) measures approximately 11 kilometres across. Note that the splashing water is suitably scaled to match the duck but would appear as giant waves hundreds of meters high in our dimension. Note how the water froze instantly – possibly not a cold freeze.

North African Shaman

Below is a picture with coordinates given in North Africa. The eye altitude set at approximately 87 kilometres depicts a side view of a human head and protruding thumb. A bird-like creature with a large beak is seen perched on top of the head, which displays a clearly visible right-sided profile of a human face showing the right eye, cheek and nose, with the rest of the face buried from the chin down. The protruding thumb does not appear proportionate with the face and may have belonged to another being or could be a result of the 3D effect.

Raw Image with slight adjustment (contrast)

Identified image auto-equalised

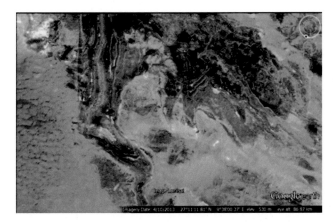
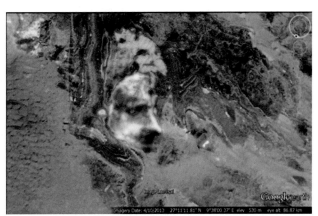

The picture on the right clearly brings out details including the creasing of the slightly-bent thumb and edge of the thumbnail. The brow, open eye and nose depicting a face are very prominent. The face measures about 25 kilometres from the top of the head (just below the perched bird) to about the position where the end of the chin might have been. It is estimated that this entity, in a standing position would have reached a height of about 187 kilometres.

Could the Earth as we know it today ever have supported such a massive being...? The answer is simply NO. Conditions concerning the solar system, including the size of the Earth, the sun and the known planets, must have been very different in order to support such huge and incomprehensible entities.

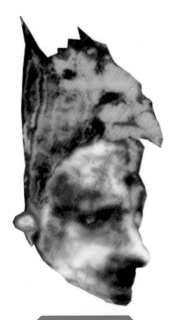
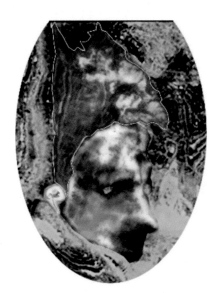

Isolating the image and auto-equalising colour and contrast brings out the detail. The bird on the head could possibly have been part of an ornate headpiece but one cannot be sure. If so however, then the features in this image reflect a ritualistic element such as that of a shaman / witchdoctor who may have been engaged in some ceremonial rite of sorts.

Bedouin

The picture below is another example of a face captured, this time in Egypt. The entity is much larger than the Shaman seen above, and they possibly did not coexist together in the same time-space in history.

Raw Image Contrast   **Saturation Adjusted**

Nothing in the image was altered apart from adjusting the contrast and colour saturation to accentuate the view of the face. Adjusting the contrast and saturation of the image can further enhance its appearance to bring out some of the hidden attributes not easily noticed otherwise.

The distance measured from the top of the eyelid to the bottom of the eye is almost three kilometres. In fact, all the pyramids of Giza would fit comfortably inside the eye of this magnificent being. The stories written of the 'gods' and 'goddesses' of ancient Egypt now have a substance of truth to them. Every story that was ever considered 'mythical' will need to be reviewed.

The head measures close to 60 kilometres across from the top to about where the chin might be. Given the average length of a human head (adult male) at about 23.9 centimetres and likewise a height of about 1.789 meters scaled up proportionately would mean that this being could have stood at a staggering height of approximately 450 kilometres!

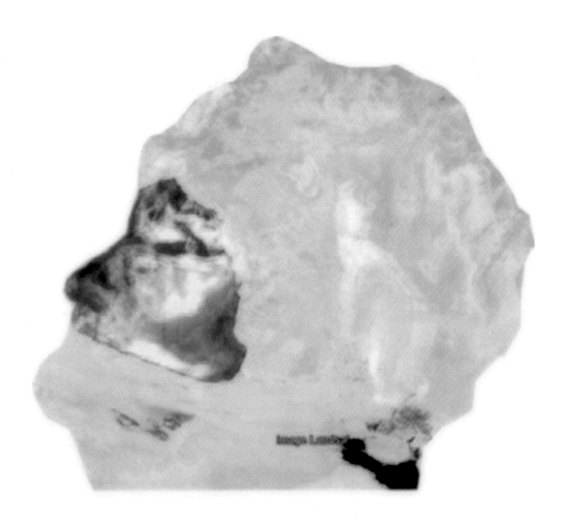

This, in our mind, would clearly seem absurd, and indeed most readers will reject this as absolute rubbish. I myself find it difficult to accept, yet the evidence is right before our very eyes.

This being could never have survived on small planet Earth... so where then did he / she reside...?? A being this size would require a powerful microscope in order to be able to see us.

At ground level, it would be totally impossible to notice any form or shape that would vaguely resemble anything like what we are seeing here. Even flying at high altitude in an aeroplane it would be impossible to make out the features of these specific beings. The perfect place to observe these entities would have to be high above the Earth to optimise the detail, which in this case would not be lower than 70–80 kilometres in altitude and looking down from outer space on a totally cloudless day.

Here at our disposal we have the perfect tool to accomplish this task, and every reader who is interested in ascertaining the truth should embark on this journey of discovery by utilising a similar protocol, uncovering this amazing world that was once hidden from our sight.

The reader should note that when observing the terrain with Google Earth, it is important to ensure that **all** special features are disabled, i.e. territorial borders, place names, cities and towns etc. This will remove all unnecessary 'noise' clouding the view and thus enabling the user to acquire a far more objective perspective of the features that comprise the terrain, where the shape and form of these incredible entities lie hidden and in fact make up the entire surface of the Earth, including every continent, mountain range, valley and sea bed.

Still not convinced............?

Namib Elephant

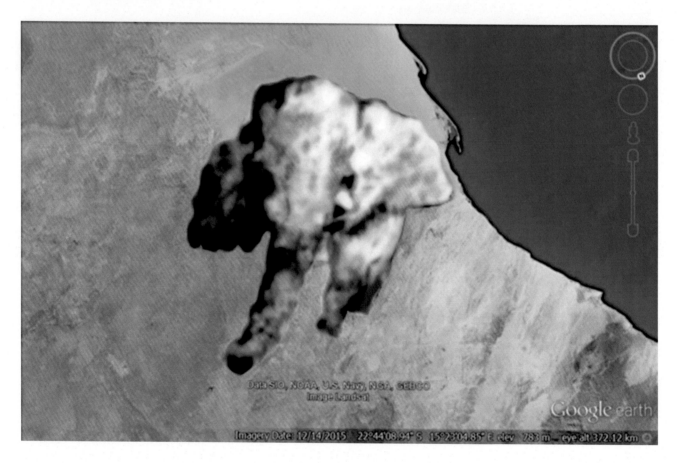

Above is a very obvious elephant that can be spotted over Namibia.

The ears, left tusk, head and trunk and both front and back legs, all clearly visible, depict an enormous elephant that once roamed around in the distant past on a planet that had a much larger surface and atmosphere.

During its lifetime, all things (flora and fauna) must have been proportionately sized to accommodate survival.

By applying a blur effect and equalising colour and contrast to the visible parts of the elephant brings out the desired detail, including the way in which the image appears to have depth and perspective as if looking at the elephant in 3D. This perspective seemingly appears on most of the images displayed within these pages. I cannot explain this phenomenon; it is a mystery.

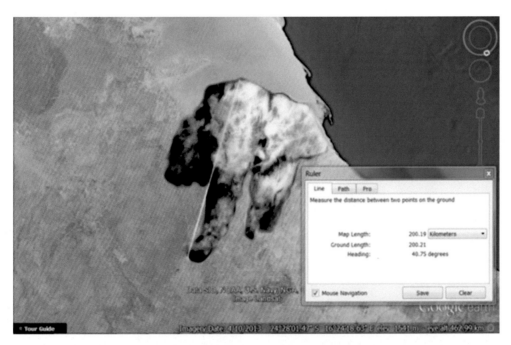

What *event* occurred that left this creature, and all the many others for that matter, in an animate posture frozen within a 3D perspective, the final second of its existence captured as if someone took a photo the very moment it occurred?

The straight yellow line marking the points between the top of the head to the bottom of the trunk measures an astounding 200 kilometres. When studying the stance and pose of these creatures, it seems to indicate that the event was sudden and totally unanticipated.

Note how the elephant was frozen in a standing position, now making up a permanent part of the flattened surface topography of our planet.

As we delve deeper into the book, it will become apparent that this elephant is a microbe compared to some of the other entities that once existed in our cosmos.

Indeed, if this elephant could stand on the Earth today, its head would reach into outer space. In other words, if we wanted to visit the top of the elephant's head, we would require a rocket to get there!

The "Warrior Monk"

Moving over to the East African coast, a much larger entity can be found that does not match the proportions of the elephant seen above.

This entity is larger in size compared to the Bedouin (page 48), measuring about 640 kilometres from head to toe.

The elephant would have stood around knee height compared with this being. The animate pose of the Warrior Monk suggests a dance or perhaps a battle pose with a shield.

Soothsayer

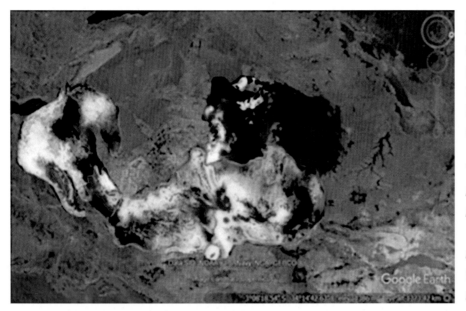

Hundreds of entities are scattered all around our planet ranging in all sizes, many resembling mythical beings we would normally consider paranormal.

This being seen on the left would stand at a height of approximately 1,900 kilometres, which is about three times taller than the Warrior Monk seen above.

Yet here presented is clear evidence that these incredible beings existed, at a certain time and in a certain space, that we have absolutely no clue about.

Did the smaller and bigger coexist together in the same time-frame? I doubt it very much. All aspects were proportionately sized to facilitate survival. This includes all vegetation, animals and living space suitably scaled to accommodate proper conditions appropriate for survival.

These entities were undoubtedly created of flesh and blood, carried in the DNA Code, just as all living matter is today.

This black and white view presents a rather 'classic' feel to this image. In many cases, the reader will question whether he / she had previously seen these things before, since many will appear familiar and even approach an archetypical view within the psyche of those who see them.

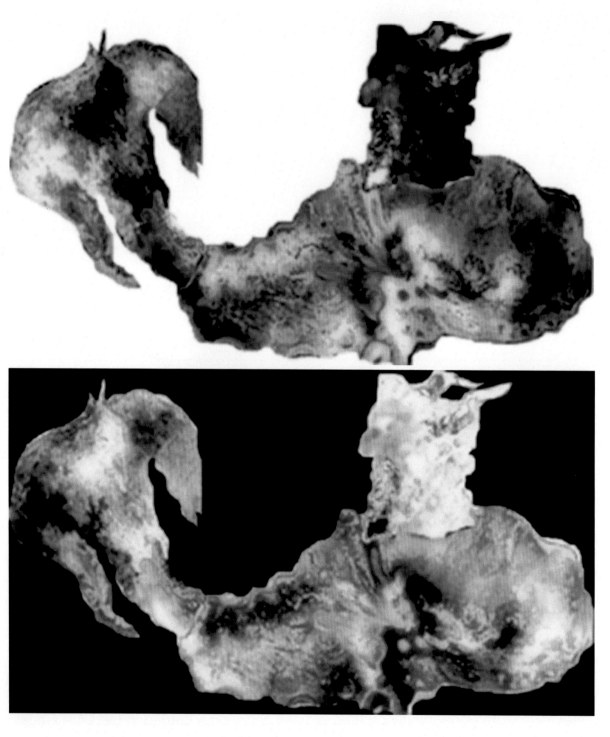

It is simply due to our own DNA recalling memories encaptured over thousands of previous generations. Many other things can be seen embedded within this image including symbols clearly noted on the sleeve of the entity in the far-right-hand portion of the picture.

Dragon Rider (West Africa)

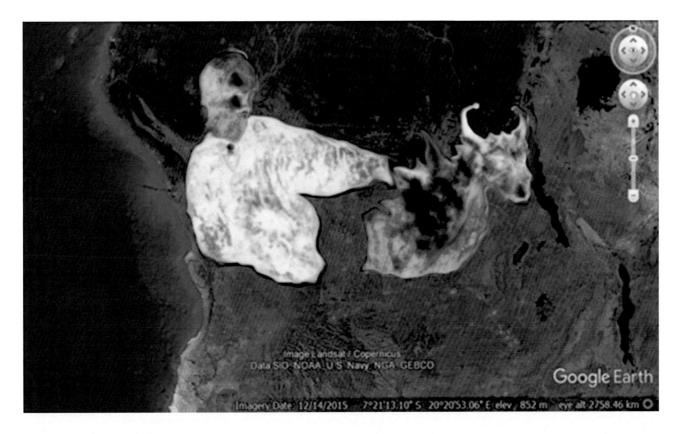

This image can be uncovered in the mid-western sector of Africa. The being would stand at a height reaching approximately 1,900 kilometres, which is about equal to the Soothsayer seen above. This means that they could have coexisted within the same space-time frame.

It is important to note that due to size differences, certain aspects could not have coexisted in the same four-dimensional system as we understand it, consisting of three spatial coordinates and one for time, in which it would have been possible to exist simultaneously with multiple-sized beings.

It would therefore be sensible to consider the possibility that there were various epochs of time, each consisting of proportionately-sized physicalities, each captured in a specific time-frame allowing all fauna and flora to exist in a suitably-sized environment.

Note that the African Rose (see pages 59–60) and Beauty and the Beast (page 81) are closely proportionate to the Soothsayer and Dragon Rider and could have coexisted in the same space-time continuum.

On the left, other filters capture a rather regal-looking being with an ornament around his neck. Possibly a leader or a king, it would have been impossible for him to have existed on Earth or in our solar system as we know it.

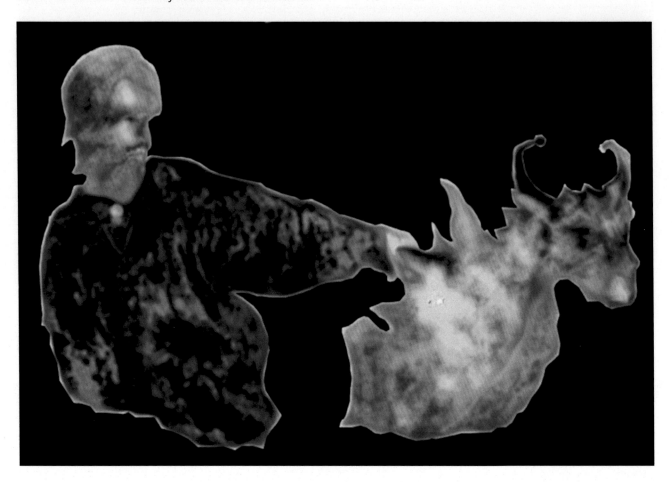

The scene above captures the exact final moment when it suddenly ended; as if a photo were taken the very second everything solidified; a sudden state of inertia that eventually turned into rock.

Memories, entrapped in our DNA, help us recall the very things that we call mythology and folklore today.

Accordingly, all things presented in this book are recognisable to the reader, as if having seen them before. We are simply recalling a memory entrapped in our DNA molecule, stemming back millions of years.

As incredible as it seems, we need to realise how we have inherited certain memories from our forefathers carried over by the DNA molecule. This is physical evidence of the Akashic Record.

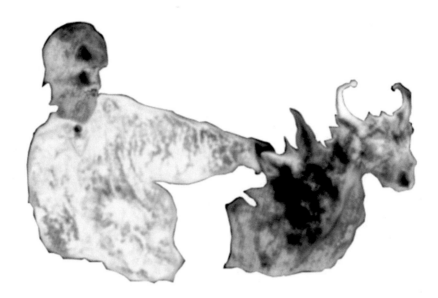

Beautiful Amazon

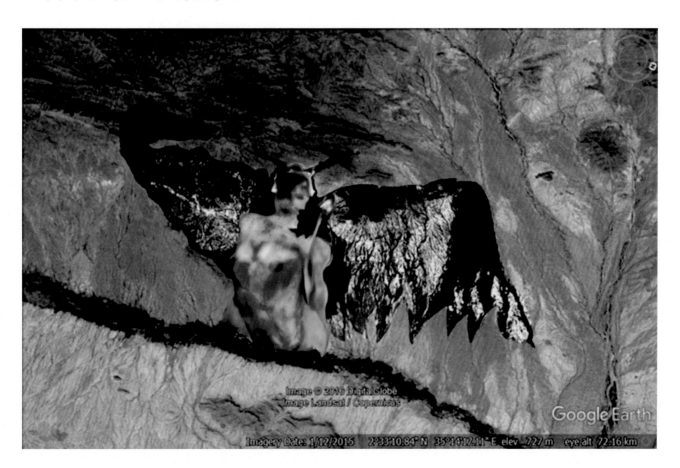

Seen here is a much smaller entity, in contrast to the others found in this region. However, if she could stand today, she would reach a height of about 40 kilometres. This image is viewed from an eye altitude of 72 kilometres.

This was a real entity that lived in a time and space we do not understand.

Mythical in appearance, she was nevertheless clearly of flesh and blood, including the fact that she had a navel (i.e. she was born) and had breasts to feed her own children.

Absolutely nothing was drawn to reveal this image. Adjusting contrast, equalising light, shadow and colour, and applying a Gaussian blur filter, uncovers this startling persona.

These beings belonged to a completely different species to humans today. Nevertheless, we see a strong resemblance to ourselves and these beings that existed in the past.

Remember, we all come from the dust of the Earth, to which our flesh will return, in this great cycle of life. Tails, wings and horns are common features. Some of these beings were beautiful, others were hideous.

Could this explain the original idea of good and evil...?

Whatever one's take on this might be, it remains an enigma pointing to an event that wipes everything out in an instant, starting a new cycle with everything, i.e. all recreated life forms proportionately resuming a smaller size than the previous epoch, a reduction possibly based on the Golden Ratio, 1.618.

This event has been repeated time and again for millennia, the evidence of which can be seen imprinted upon the face of the Earth, and, as a matter of fact, all over the universe (see Chapter 8).

Clearly, the evidence suggests that to support such gigantic entities, the habitat they existed in must have had a greater surface and atmosphere compared to today, or alternatively these creatures existed in an ethereal setting we cannot explain.

Garden of Eden

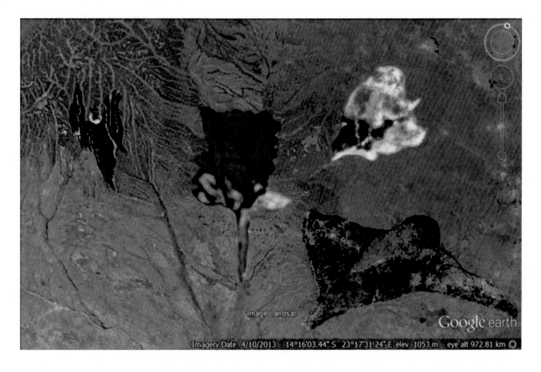

On the left is an example of an enormous rose that can be found in Central Africa. The section of rose and stem that can be seen is bigger than the Namibian Elephant. Once the image is captured, it is easy to adjust the contrast to bring out the true colour of the rose and stem.

The rose measures more than 360 kilometres from top of petals to the bottom of the visible stem (green section). This will also give you an idea how large the other creatures are, that can be seen in this picture.

Among other things highlighted are a bird's head, a winged being and a fairy-like creature hidden behind a bunch of flowers.

Could this reflect a type of Garden of Eden... repeated accordingly within the Fractal...?

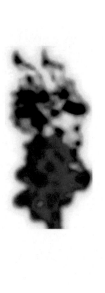

The section of the rose and stem is larger than the Namibian Elephant by approximately 150 kilometres in height.

Each epoch came to a sudden ending, leaving behind an imprint, like a watermark noticeable to the keen eye. Each new beginning then starts over, assuming a smaller size preceding the former event. This evidence is measurable and appears to meet the Fibonacci sequence, based on the Golden Ratio.

Most of the landscapes selected for the book have been largely undeveloped for thousands of years. Here, within these pages, best visibility is obtained providing a better base to indicate how these entities, embedded together in the topography of the earth, are comprised of multiple-sized aspects that may have existed in different space-time cycles in the past.

Clearly, if one accepts the 'cycle' theory, then the larger beings and creatures must have existed further back in history and the smaller in more recent times.

The Sinai Desert.

Hundreds of entities can be observed scattered all over the Sinai Desert region ranging in all sizes. Various areas have been highlighted by equalising shadow, light and colour.

The following are viewed from an eye-altitude of about 75 kilometres. Carrying out the above adjustments brings out the desired detail. The evil-looking being with the hood measures about 26 kilometres from the top of its hood down to its waist. Many other beings can be noticed even without any adjustments necessary.

Note the clarity of the eye on the face below, preserved from erosion

Contrast only   **Equalised and Coloured**

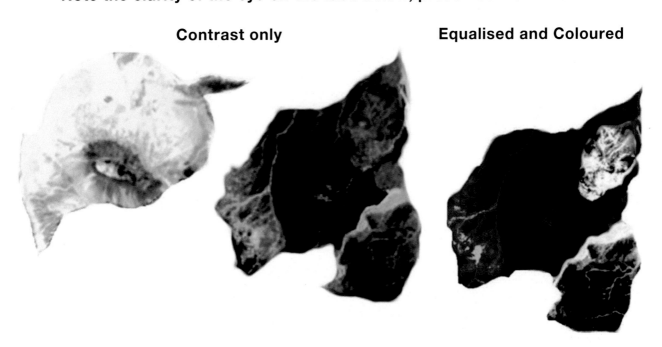

The daunting features of this entity seen above are clearly intimidating, while holding what appears to be a book or tablet of sorts in its left hand.

"Moses" and the Ten Commandments...??

Zooming at a higher altitude (238 kilometres) reveals a striking image dominating the central point of the Sinai, reminding one of Moses and the Ten Commandments, especially since this event occurred in this exact location where Moses, according to the Book of Exodus, received the Commandments from God on stone tablets. Many other features can also be seen, such as the much smaller priest-like entity dressed in a robe.

What is also noticeable is the eye in a triangular setting on the 'tablet'. Could this represent the all-seeing eye of God as depicted by the ancient Egyptians? This 'stone tablet' measures about 100 kilometres in length and the entity would have stood at a height of about 300 kilometres.

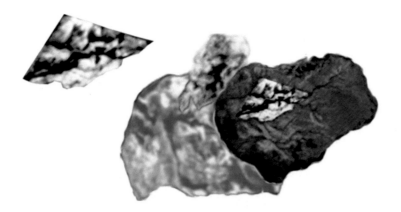

Is the story of Moses, as we have been led to believe it, perhaps a repeat of a much older event?

Below is a view of the Sinai and Arabia from a different angle and altitude, with Egypt in the top right-hand corner above. When looking at the raw landscape it is difficult to notice anything unusual which may be of interest.

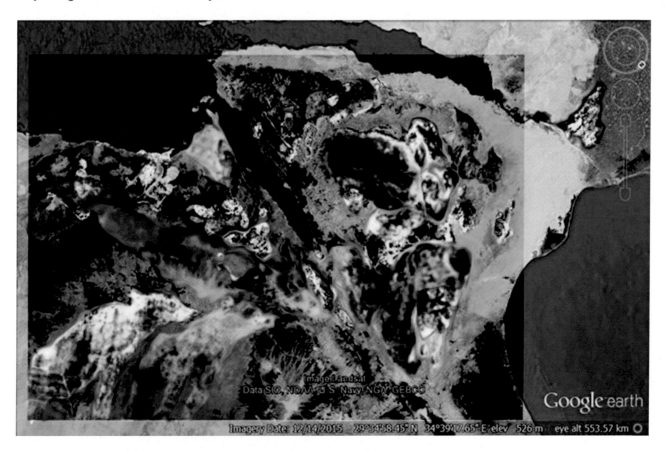

However, by darkening a segment and adjusting contrast will bring out features that are rather amazing. This is a quick way to identify any unusual features, and then to extrapolate using the relevant techniques.

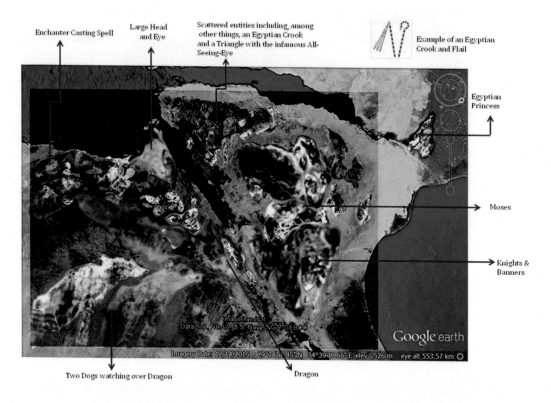

Typically, the terrain consists of multiple characteristics ranging in different sizes forming the topography and geography of the landscape. Careful application of the tools and filters identifies hidden features, bringing out all the detail noted in the pictures presented. The cartoon-like appearance of the image below is purely a result of the filtering process applied to highlight the detail. No structure is added to the image, which clearly depicts some kind of sorcerer holding a ladle with a dragon emerging out of it. *What does all this mean...?*

Note the design on the wizard's hat, which appears with a badge and inscription. A strange-looking entity appears close to his left hand.

Other features such as this goblin, together with what appears to be a bird's head, can be found on the right side of the bend on the crook on the south-eastern portion of the Sinai.

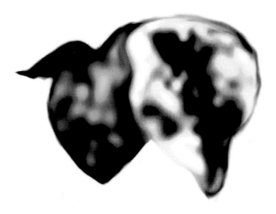

Knights and Banners
What could this symbol mean?

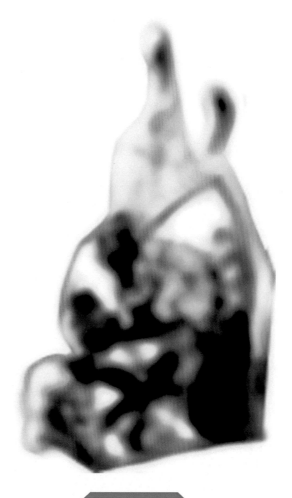

"*Moses*" viewed from a different altitude and angle.

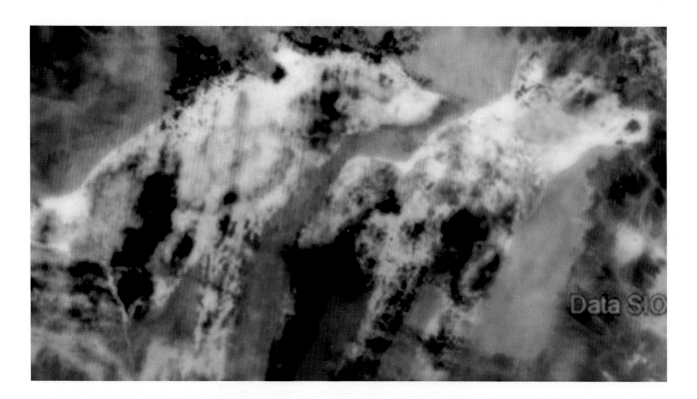

Two Hounds can be observed as if in 3d

Life forms of all sizes can be noticed scattered all over the terrain, which is largely a desert and has remained undisturbed for thousands of years.

The detail within this picture is obtained by simply utilising the Gaussian blur technique. This shows how the 3D effect appears to be embedded, almost like a watermark, but at the same time forming the topography of the surface. Higher blur adjustments bring out other images embedded deeper and in-between.

Here rests a mystery beyond belief, previously hidden from the human eye. How far back into history are we looking at??

Apart from our understanding of mythology and folklore, the true history of the world is very different to what we have ever been taught. Revealing the substance of the reality that existed may not be anything close to how we may have imagined it to be, thereby rewriting our history books and everything we have been taught to believe.

A mix of various Characters, some appear Dark, others like Angels

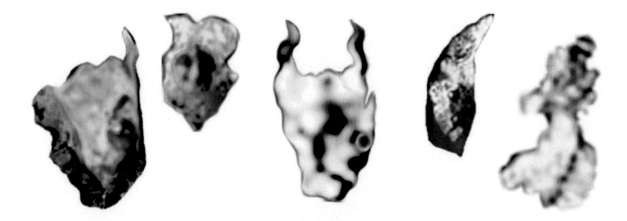

A world filled with all kinds of life forms – angels and devils, demons and wizards, fairies, witches and even babies are noted... a fantasy world in every respect.

Once again, this may show how recognition is possibly captured and memorised through the human DNA and carried over thousands of generations, like the Egyptian crook seen on page 64.

It also explains why there is a certain familiarity when observing these beings and creatures, as if we have seen them before. Countless artists have expressed similar representations in fantasy and mythical art.

Amazon Basin Entities

The very things encaptured within our consciousness determine the very nature of history as it would play out its course as encoded in the human DNA molecule. Here, among many other things, can be seen what appears to be a sage-like character with numerous sacred symbols inscribed on his clothing. What role did this entity possibly play in the past?

And what does the large spot at the centre of his forehead represent? …. Could it be the third eye…??

Aside from what appear to be magic symbols inscribed on his clothes, many other aspects can also be captured, such as the fairy-like creature visible just above his staff.

One can also notice a warrior-like being with a shield just to the right of the highlighted segment; like that used by the ancient Greeks. This image is captured over the Amazon Basin reflecting civilisations that existed millions, if not hundreds of millions of years ago. Does history repeat itself??

Amazon Warrior   **Fairy-like Creature**

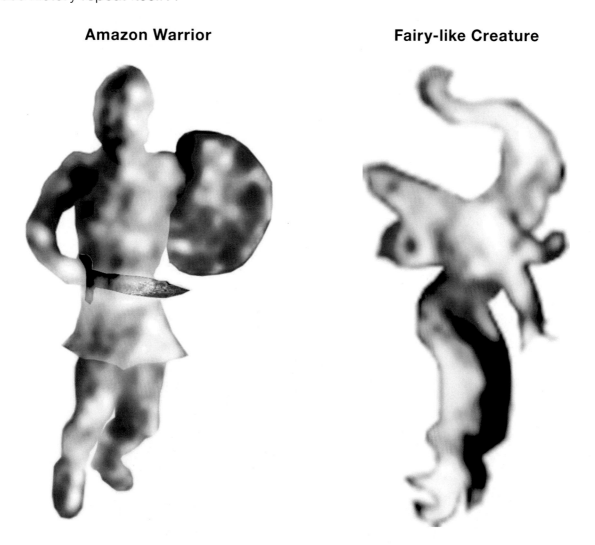

Equalising and inverting the colour values brings out the detail seen on the sword.

Sabrina the Witch

Clearly a view reminiscent of a witch, note how this image correlates with how an artist might typically portray a witch on a broom.

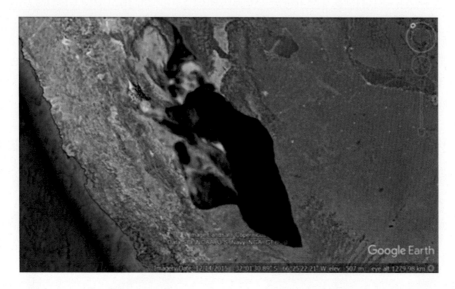

Perhaps by adding a full moon to the image this may explain a transcendent secret, as many fairy tale and fantasy art depictions of witches riding brooms have a full moon in the background.

Here, to demonstrate the reality, the same image has been superimposed against the backdrop of a full moon. Were these real witches who lived in a world very different to our own today?

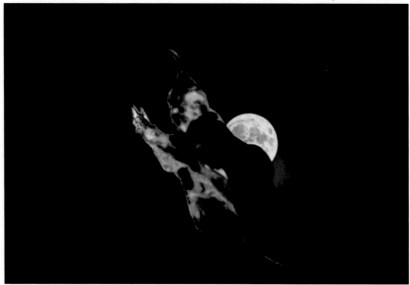

Raw Image background source: https://cdn.pixabay.com

Perhaps certain spells were cast in past times that still captivate the human consciousness in believing lies rather than the truth.

If so, then these powerful spells holding humanity captive in a great deception are about to be undone... and hidden things will finally be revealed.

This is the great awakening, the Apocalypse.

THE ANGELS

Everything is caught unawares in a flash of time.
Even Angels did not escape this Event!

Angel with a Message.

Note the Symbols
on the Scroll ←

WELCOME TO THE 5TH DIMENSION

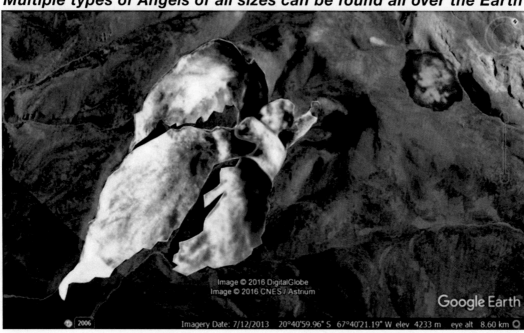

Multiple types of Angels of all sizes can be found all over the Earth

The angel above is holding a scroll and seems to be leaning towards a portal that I decided not to uncover. However, the reader is encouraged to investigate further at your own discretion.

Angel with Scroll

This angel appears to have been in flight at the moment it all ended. What does it all mean?

This 'Beautiful Angel' can be found inside the old delta leading into the Gulf of Aden about 265 kilometres from Djibouti. The angel also appears to be in flight.

Contrast added to detail

Raw image

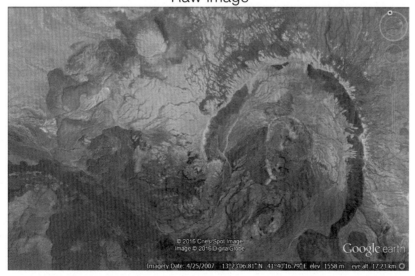

What is very interesting about this image is the infant-like angel (Little Wing – see below) holding onto the larger angel who appears to be nursing this child.

This may serve as confirmation that the angels themselves proliferated and were flesh and blood just as we are. However, they are certainly a different species, possessing attributes above human capability, such as winged flight. In some cases, tails and horns are also noted, although most features are of human qualities.

This angel would have stood approximately 5.08 kilometres tall.

Beautiful Angel

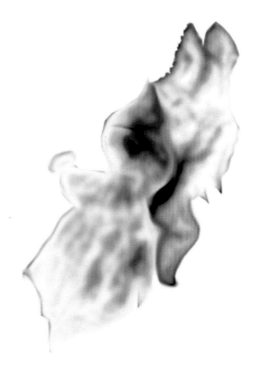

Little Wing

Note the detail of this infant, including its outfit and the small wings. The pose also suggests breastfeeding.

Ice Angel

The 'Ice Angel' is one of many angelic beings found on the Antarctic, which is Earth's southern-most continent and the geographic South Pole.

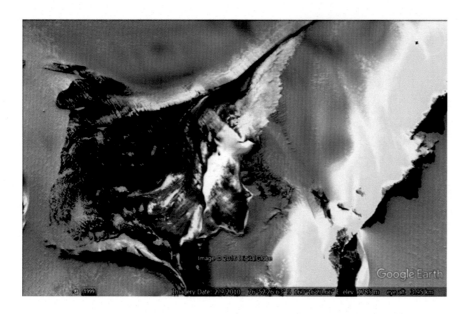

On average, Antarctica is the coldest, driest and windiest continent and has the highest average elevation of all the continents.

Antarctica is considered a desert, with an annual precipitation of only 200 mm (8 in) along the coast and far less inland.

The temperature in Antarctica has reached −89.2 °C (−128.6 °F), though the average for the third quarter (the coldest part of the year) is −63 °C (−81 °F).

There are no permanent human residents in Antarctica, but anywhere from 1,000 to 5,000 people reside throughout the year at the various research stations scattered across the continent. The question is… was this continent located in a different position on the Earth during the lifetime of this angel – or is it simply a case of our planet having a very different climate and topography in times gone past?

The Ice Angel measures approximately 840 meters in length head to foot.

Again, notice the rather animate features of this entity, suggesting that it was in motion at the very moment it all ended.

Were the angels themselves unaware of this event; this sudden occurrence immobilising all matter into a petrified state, solidifying and eventually turning into rock?

Ancient Angel

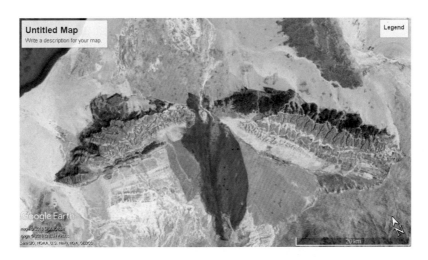

This angel can be located in the Horn of Africa at coordinates 10° 20'38.46" N 45° 21'35.63" E, then adjust eye altitude to 60 kilometres and a north direction as noted in the image above.

What function did this angel perhaps serve? Was he an emissary from God...? And if so, why did he perish like all other organic matter in an instant of time??

His body length spans approximately 22 km across, measuring head to foot.

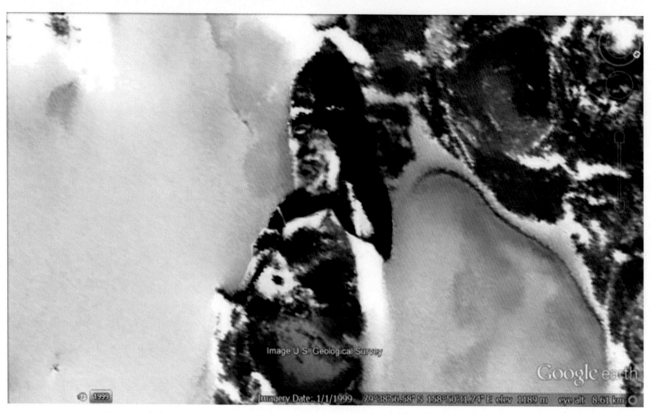

Antarctic Priest

Here we see the unmistakable features of a priest, dressed in a robe, hood and tassel and seemingly standing behind a pulpit. If he were alive today, he would stand at a height of approximately 7.3 kilometres.

Why was there a need for an apparent ecclesiastic millions of years ago?

Was there possibly also a type of 'church age' or 'religion' of sorts during this time?

Is this a typical depiction of the fractal pattern repeating itself??

**Of what order did this celebrant belong... and
what was the message he preached?**

The Antarctic Ghoul

This disturbing image resembling something like the Grim Reaper (without the scythe) can be found on the Antarctic. Clearly, here is a being normally depicted in **mythology** and **fables** appearing as a dreadful phantom holding a staff with a serpent slung over it. What part of this history was eradicated out of the human consciousness, and who is responsible for keeping the human race under the spell of a great deception?

Raw image	Coloured to bring out detail

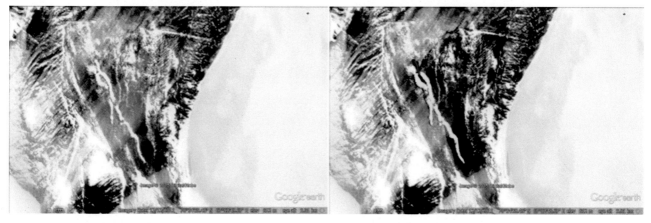

It is clearly time for the human race to stand up and to shake off the chains of bondage that keep us bound to the deception that is ruling our planet. This information should give each and every normal person a wake-up call that we have been manipulated left, right and centre to the tune of the **ruling elite**, who have possibly hidden this knowledge from the eyes of humanity – or perhaps they simply do not know about it at all...?

The Ghoul is about 3.7 kilometres long.

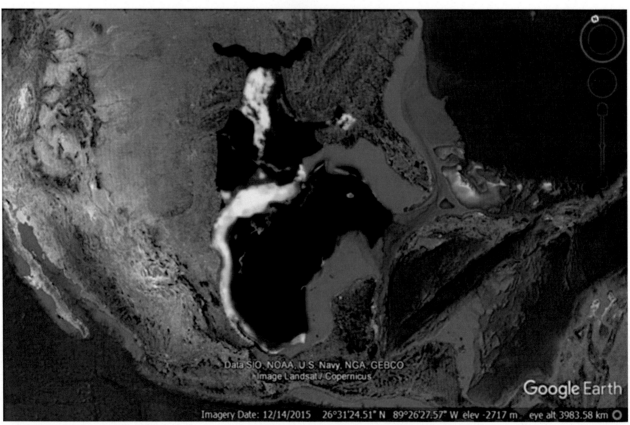

Beauty and the Beast

Standing height about 1838 kilometres.

Here we have an incredible view of a being who appears to be riding on a mythical beast, perhaps a type of bull or dragon.

A certain event occurred in the past that caused this being and beast to suddenly crystalise. The same characteristic is seen in all of the images featured within these pages, therefore it is worth assuming that this will occur again in the future; an event so sudden that no-one knows when it will happen.

Does this event have something to do with being exposed to a galactic wave of photons in cycles of time, determined in a dimension governed by a fractal? These and many other questions remain unanswered…

CHAPTER 5

ZEITGEIST

Witches and Wizards

This chapter presents further evidence and a view of a similar culture existing further back in the past. Note how these aspects had adopted a set of ideals and beliefs predominantly expressed today in mythology and religion. This consciousness is passed down over time via the DNA molecule that we inherit from our forefathers, going back billions of years. Our history is moulded within our DNA as a result of the lifestyle adopted by the life forms that existed, including all the enchanters, angels, witches and wizards noted here. Perhaps this is how we receive our genes pertaining to good and evil – though some would say these attributes are spiritual in nature. Certainly something to consider...

This image is captured by inverting the colours to help identify additional hidden features. The above can be found as per the coordinates seen in the frame at the same eye altitude and compass orientation. Here, several beings of a mythical appearance, including a wizard-like being, take up an enormous part of the Earth located between North and South America, with both continents resting on his shoulders.

In the image above, a spirit-like being seems to be whispering into the wizard's ear, while another appears to be eavesdropping. Symbols and other inscriptions can be seen on his chest. Keep in mind that this is the rock of which the Earth is formed.

Below, a much smaller priest-like being is noticed holding on to what appears to be a scroll. Alongside to the right is a larger man holding a baby-like entity dressed in a strange outfit.

Below is what appears to be an enchantress, holding a wand in her right hand and casting a spell, or perhaps conducting music?

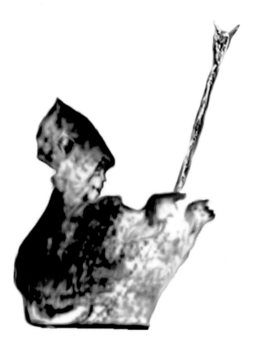

She seems to have a bonnet on her head with a jewel-like feature embedded in the centre.

To appreciate the size of these beings, consideration must be given to the viewing distance, which in this case is approximately 10,297 kilometres from the surface of the Earth.

Magic and sorcery appear to have been deeply embedded within the lifestyle of most of these entities discovered.

The Mystical Past

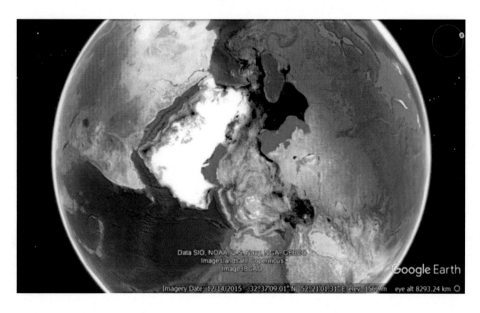

This easily recognisable view of the above can be captured by carefully identifying the attributes making up this character. The Entity is holding a Trident.

The Sentinel

This being measures about 3,873 kilometres in height and existed when the solar system, as we know it now, had not yet formed. Again, this view epitomises how the reality of our past is more akin to myth and fantasy than we were ever taught.

The fairy tales we tell our children are without a doubt based on a reality that once existed in the past.

This is captured in the DNA molecule from which all humanity stems and is subsequently passed down thousands and perhaps millions of generations and is now manifested as a memory within the consciousness of humankind.

Note the battle pose, as if about to strike a blow…. The Arabian Peninsula is his shield.

Mother Goose's Brother

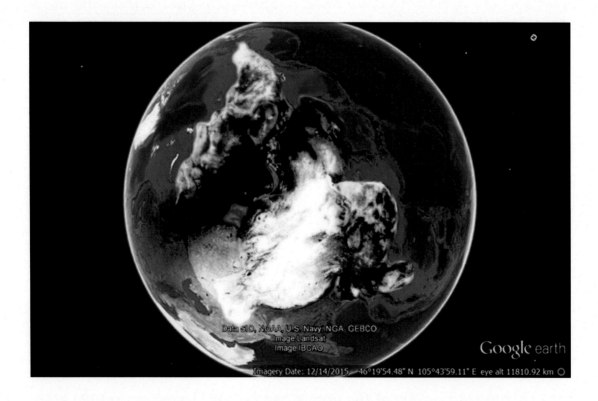

This feature covers most of the Asian Continent. A single year during the life span of this goose would amount to millions of years for a human on our planet today.

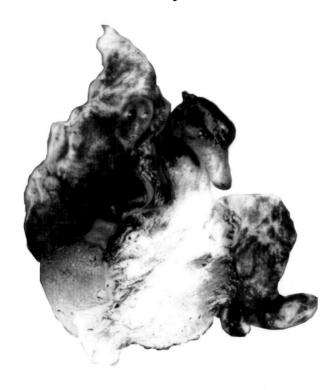

This is the reality of the world we live on.

What is striking is the fact that most of the features presented appear as if posing for a photograph, with absolutely no indication of a pending event that freezes everything in an instant.

The goose alone measures 6,500 kilometres across in a straight line from the top of the head to about the end of the tail.

Could anybody out there provide a possible answer to perhaps the greatest mystery ever to be contemplated...?

Mother Earth

An incredible view of a woman who existed perhaps billions of years ago, note that the breasts and facial features seem to indicate that this was a young woman. The question is... where did all these huge beings and creatures reside?

It would seem reasonable to conclude that 'Mother Earth' (the name I chose for this being) must have existed on a much, much larger planet long before the solar system was formed. This could possibly be traced by measuring the size reduction ratio and correlating the data to identify prehistoric space-time dimensions (see previous chapter). We are clearly looking at a period when the space-time cycle within the history of the fractal was much larger than it is today. However, this being is still minute when compared to the larger components of the fractal, as noted when observing constellations in our Milky Way galaxy.

Mother Earth would have stood approximately 26000 kilometres tall.

The Dark Emissary

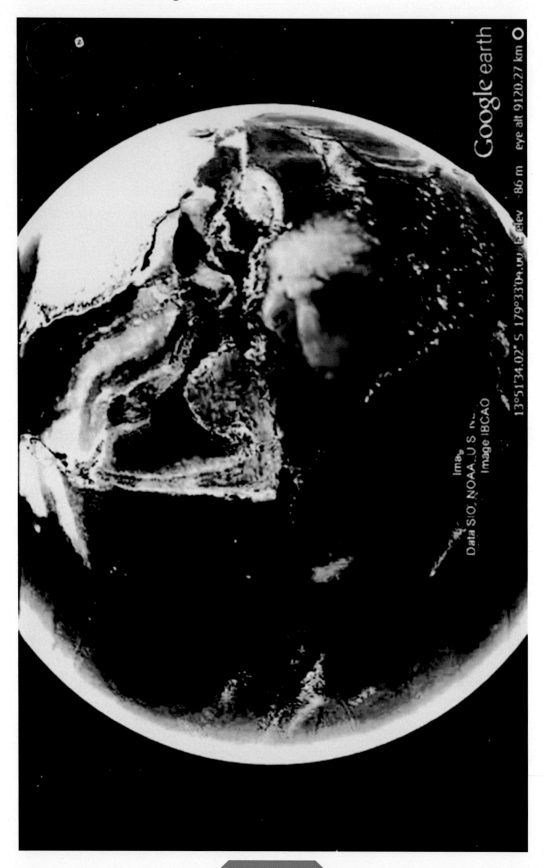

The Dark Emissary

**The truth is more unbelievable than what
an utter lie will ever be!**

Here is yet another example of an enormous being found as per the coordinates given above. Note the face of this dark-looking entity seemingly peering from a cloak (forehead, eyes and nose) with a beret-like headdress sporting a golden badge with a little flag attached.

A golden chain with visible links hanging over the forehead can be seen attached to the badge. One horn is clearly visible.

He appears to have held a high position taking the headdress into consideration.

Are these perhaps the 'Rulers', the 'Principalities and Powers' that are spoken about in the Bible?

Standing, this being would reach a height of approximately 30,800 kilometres. An entity this size, living today would suck all the air out of the Earth's atmosphere in a single gasp!

A good athlete nowadays could complete a 100-meter sprint in about 10 seconds. However, if scaled to the size of the Dark Emissary, an athlete could possibly complete approximately 1,721,632 kilometres in a mere 10 seconds on our current space-time scale, meaning that from our perspective he could run at 172,163 kilometres per second, which is about 58% of the speed of light. It must be noted that the bigger beings could easily exceed the speed of light as we know it in our dimension.

An environment scaled to suit this entity consisting of a far bigger Earth, and bigger solar system scaled proportionately, would subsequently show that a single day would measure 47 years on our scale. It is obvious that the solar system today could never have accommodated such a massive entity. Clearly, even the smaller beings shown in the previous chapters must have dwelt in a completely different environment, and also could not have coexisted together within the same dimension of space.

All the matter making up planet earth and the other planets are the remnants of these aspects that could have been subjected to the Photon Belt phenomenon. This cycle is repeated over billions of years and the time scale governing it is within the fractal space it resides in accordingly.

We are clearly observing multiple cycles of space-time that we are yet to explain. The larger aspects appear to have survived further back in the past and the smaller varieties in more recent times. However, this is just a theory at this stage, and may well prove otherwise; for example, different dimensions, all coexisting at the same time, but entirely independent from each other.

Whichever way you wish to believe, it remains clear that the formation of the solar system was never formed in the manner we have been taught. Based on the findings presented herein, cosmology, anthropology, religion and history will all come under serious review. Our current political ideologies will be changed and every human on earth will realise we are but MICROBES resident on a planet that possesses a history we have absolutely no idea about.

Furthermore, these findings prove that the earth is made up of the remains of life forms that once existed possibly billions of years ago, all having been overcome by some instantaneous event which froze them to eventually fossilise into rock and later provide the substance for survival for the next cycle of creation. This may explain the process of the entire universe.

Planets are the substance of compressed matter that was once organic. Further compression heats the central core of a planet until it vaporises under pressure and becomes a star. The earth may need to be compressed to the size of a football before it might become hot enough to vaporise and be considered a star. Taking this further, the sun may well have been a huge planet originally that was once comprised of organic matter compressed over billions of years and eventually becoming a star.

Nuclear Shadows

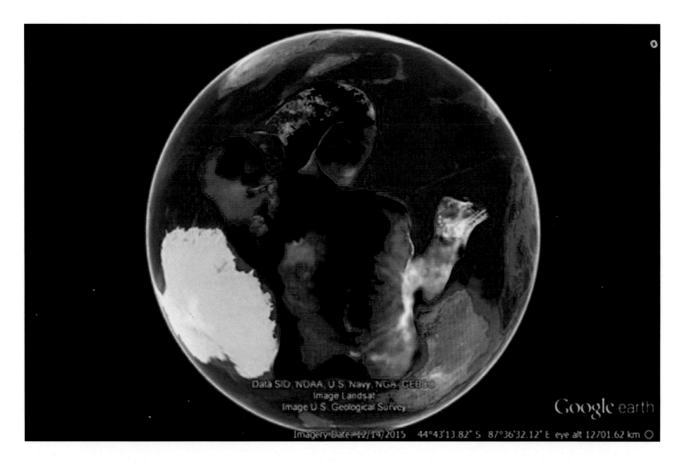

Permanent shadows, also called 'Nuclear shadows' were formed during the bombing of Hiroshima and Nagasaki. Could a sudden emergence in pure photon light produce an after-effect such as this...?

This feature shows a being holding an item in the left hand and another item slung behind the back with a cord attached to the right hand. I cannot explain the shadows that are cast, especially the cord slung over the back. It is a mystery to me.

Shadows are noticed on most of the features presented

Nuclear shadows...?

Perhaps this might provide a
clue to what the 'Event'
may be.

Canadian Hare

The images presented in this book are going to force evolutionists to reconsider their idea that ALL organic life evolved from a single random event. Can something as complex and wondrous as the natural world be explained by a simple theory? Clearly now, this cannot be the case.

We are all very complex creatures, and this includes planet earth itself. The planet we live on is a responsive, pulsating (electromagnetic field) and living consciousness. If you don't believe this, then try living without animals, plants and minerals – all contributions from the earth. Everything we require to sustain life comes from this same life, including the air we breathe, the soil we walk upon and the water we drink.

As can now be plainly observed, there were many creations that could possibly have occurred. It now becomes totally improbable to suggest that evolution is correct as currently taught. This hare measures an incredible 3,240 kilometres across from the tip of the left paw to the tip of the tail. Even the texture of its fur is still noticeable.

South African Satyr

Here is another example of an enormous entity. The image is that of a face of an impish-looking being that existed perhaps billions upon billions of years ago. Note that the bottom part of the jaw is no longer visible.

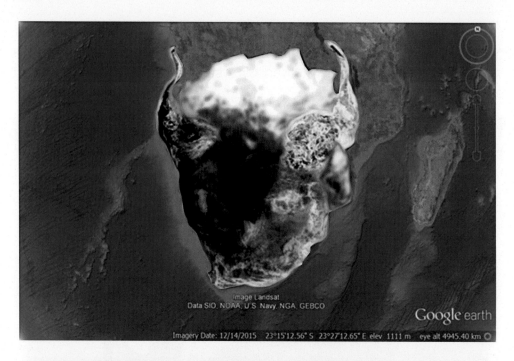

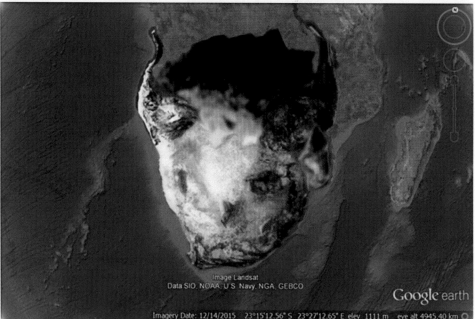

Above is a view of the Southern African region, which in this case has been highlighted by applying automatic contrast (auto-equalise, Gaussian blur etc.) No colour has been added to the image apart from the eye, visible in light blue. The image (head only) on the top has its colours inverted to highlight further detail.

Notice that the teeth (top and bottom row), where the upper lip would be, are still visible. Part of the eyeball on the right is very noticeable and the detail of the ear on the right is still discernible.

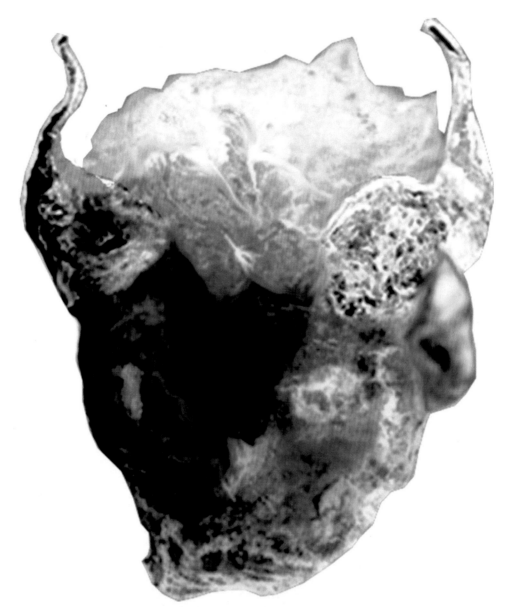

Inhabitants of Southern Africa: Behold Your Dwelling Place

This image viewed from approximately 5,000 kilometres clearly shows that the entire southern part of the African continent is the remnant of a males' head, which very clearly takes the form of a mythical-looking being.

Two horns are noted, and a beard that now makes up the Drakensberg Mountain Range is still visible.

If this entity could shed a single tear from the eye on the right, it would result in a flood of Biblical proportions that would wipe the entire Province of Gauteng in South Africa off the face of the map, including every living thing in the path of the tear flowing down his cheek. Indeed folks, we earthlings are the dust of the earth.

We are in fact microbes by comparison to what existed before. This bearded being would have stood at a height of about 19,000 kilometres.

The remains of the smaller creatures can be found upon these larger entities, including all the flora and fauna.

Clearly, it would have been impossible for all these things to have existed collectively during the same period, as the bigger could not have survived simultaneously with the smaller features, unless each was nested in separate dimensions. Does this perhaps indicate that there have been multiple creations taking place in cyclic time, and which may continue into the infinite future?

Does a regular event both end and start a new epoch in continuum with fractal time? Clearly, the evidence of such events is noted all over the earth. The next event is therefore as inevitable as it is sudden, it will occur without warning, and just like in the past, nobody will be prepared for it.

Why is everything in our dimension so much smaller than these things that existed in the past? Everything found can be arranged in differing sizes, which therefore begs the question: is everything assuming a proportionately smaller dimension, specifically following an aftermath, when everything is recreated on a smaller scale meeting the noted regularity of a pattern as depicted through the Golden Ratio, 1.618? (See pages 134–135.)

All the images depict beings and creatures captured in a state of sudden inertness, frozen at that very moment in time, as you see them within these pages. Could these entities have been exposed to a sudden burst of pure photon light, which at that moment may have caused all atoms to cease vibrating (becoming motionless), making everything 'freeze' in a moment of time? And could this possibly have something to do with the fact that photons have no mass?

In all the cases that have been discovered thus far, all these aspects, including the angels do not show any sign of hiding from a pending disaster. All seem to be captured candidly at the moment of a sudden and unexpected event.

CHAPTER 6

ARTEFACTS

Huge folded cloth-like fabric (South America)

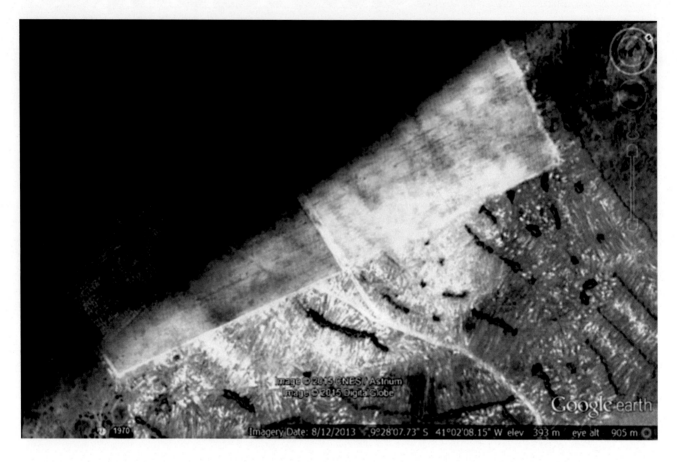

The picture above is found in the Amazon region of South America.

It clearly shows a section of folded fabric that appears to have been pulled up and folded over part of the terrain. The physical attributes are clearly cloth-like, depicting a frayed or cut appearance visible on the right side of the folded material.

Furthermore, the torn appearance of the part of material that was stuck to the surface of the Earth is still visible on the upper left part of the folded material.

Is this perhaps part of a garment??

The following images show finer details of this fabric wonder. Note that each stitch notch is about the size of a small passenger vehicle.

A zoomed in view reveals the detail of the weave pattern.

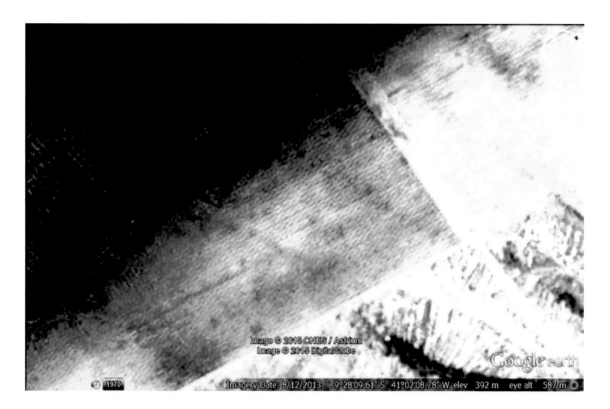

These images were not subjected to any of the techniques mentioned in this book and are exactly how this would appear on Google Earth (as per imagery date). Note the frayed appearance of the edge of the fabric showing its cloth-like properties.

Below is another view of the detail of the woven fabric and how the covering was pulled up from the ground leaving a torn appearance at the base. Notice also the stitching clearly visible and perfectly manufactured. This is hardcore evidence of a huge piece of fabric-like material that must have been manufactured on a gigantic machine perhaps millions of years ago.

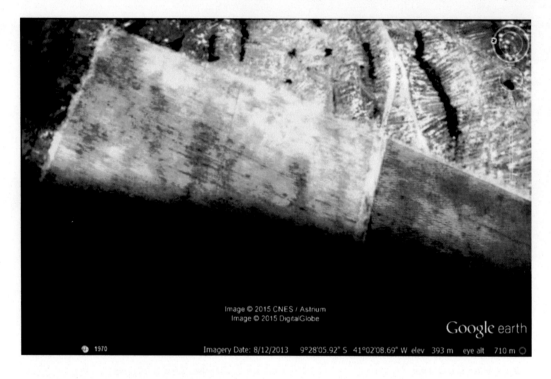

Book with Visible Text

Part of a manuscript perhaps. Here we see the frayed pages and text that is still visible from approximately 568 meters away. The frayed edges can be clearly seen on the right-hand side, revealing a manuscript older than we could ever imagine.

The left-hand page seems to bear a design, like that found in a Middle Age Manuscript.

Perhaps someone could help decipher this Sacred Text..... ?

This text, still clearly visible, must have been written millions of years ago by some enormous entity.

It would be most interesting to know what it says. A message from the distant mystical past perhaps?

Colour Inverted

CHAPTER 7

THE SOLAR SYSTEM

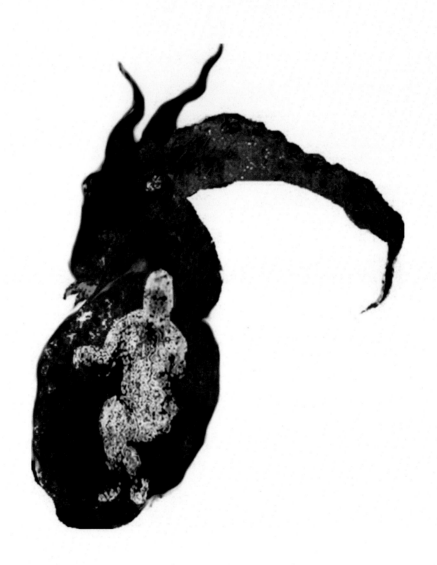

The Solar System, showing Pluto just inside the Kuiper Belt stretching almost 5 billion Kilometres from the Earth.

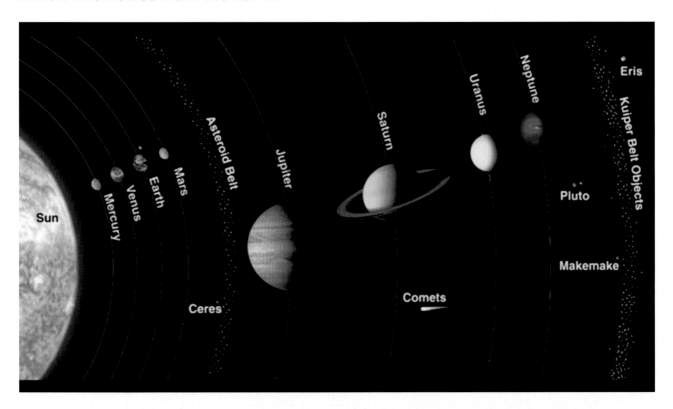

Image Source: https://i.pinimg.com/originals/6f/6d/09/6f6d09d4c7dd69c814745dc7e1bd4bdf.jpg

Due to the fact that the following images were not obtained via Google Earth, I am unable to provide coordinates for the reader to check their validity. However, the reader may easily obtain authentic images of any planet off the internet (courtesy of NASA, in spite of some of their cover-ups), and all the images displayed in this chapter can be found accordingly.

To prove my point that all the rocky planets in the solar system were formed in the same way as can be seen on Earth, I have examined the furthest planet from the sun, namely Pluto, of which we have obtained reasonable photographs. Although no longer recognised as a planet, Pluto and its largest moon, Charon, astonishingly reveal the same aspects in this distant, desolate place some 5 billion kilometres away from the earth!

NASA's New Horizons' probe, launched in January 2006, flew past Pluto after a voyage spanning some 4.8 billion kilometres. Leaving Earth at 57,600 kilometres per hour (36,000 miles per hour) and covering about 1.4 million kilometres per day throughout its odyssey, it had still taken the spacecraft nine and a half years to reach its target. New Horizons streaked past Pluto in July 2015, and pictures and other data were beamed back to Earth with a pair of 12-watt radio transmitters. It took those signals, traveling at 297,600 kilometres per second, 4 hours and 25 minutes to cross the solar system to our planet.

The following pictures were published by NASA in 2015 during the fly-by of the New Horizons' probe. I subjected some of these photographs to the same filtering techniques, i.e. equalising shadow, light and colour, and in some cases applying a suitable blur technique to defuse fragmentation etc., bringing out the true nature of what is hidden in the raw images. Some of the images have been rotated to obtain additional features. Readers are encouraged to engage in this technique before dismissing these findings as nonsence.

Some readers may find the images in this section of the book rather disturbing. These images are not drawings / artwork but were uncovered by means of the same simple technique as was applied to uncover the detail on Earth.

Dragons on Pluto

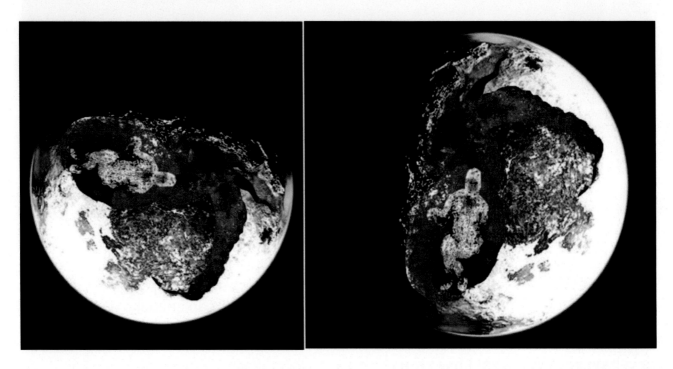

The dragon (coloured green) can be clearly seen in the top left-hand photograph above. Rotating the picture counter-clockwise by 90° reveals another dragon (red) with a being embedded together.

Again, it is necessary to emphasise that these images were not drawn. The raw image was subjected to the same techniques as employed with the images obtained off Google Earth.

The 'Mythical Dragon' is real and inhabited the early universe in a time-space we are yet to understand.

Captain Pluto

Raw image
Image Source: https://blog.mordorintelligence.com

Image with filters applied

NASA's New Horizons' fly-by has unwittingly revealed a greater mystery about the dwarf planet than what they bargained for. The US space agency released the above image of Pluto and its moons. The image above (Pluto alone), subjected to the same filtering technique, reveals an astonishing view of an entity, which I named 'Captain Pluto', seemingly gazing at Charon, the planet's biggest moon.

It is interesting to note that the New Horizons' team refers to a vast dark region covering an area of over 320 kilometres in diameter on Pluto's north pole as 'Mordor', from *Lord of the Rings* fame.

Notice all the other beings seen in the filtered segment of Pluto, all of which indicate the widespread reality of the substance of every planet in the solar system. Hundreds of entities can be uncovered on both Pluto and its moons.

Organic matter, which existed in the past, is compacted layer upon layer forming the topography of every rocky planet in existence. This is exactly how the Earth is structured.

As can be noted in the picture above, this being could simply not have lived on Pluto as we know it today.

Pluto, and possibly all the planets of our solar system, could be the remnant of a much larger planet that existed in the past, in a much larger configuration pertaining to the solar system.

In the same picture, other beings (some much smaller) are noticed scattered all over the planet.

'Captain Pluto may appear like a storybook character, however, the fact remains that this being, whatever role he played as a living entity, must have resided in a very different cosmos totally unknown to the human race. Along with him were myriads of other beings and creatures, some dark and others angelic.

DARK Planet Pluto

All of the raw images of Pluto and its moons were taken by NASA's New Horizons' probe. By simply applying the same technique as defined in this book to any of the raw photographic images presented by NASA, this will show the absolute proof that the entire solar system includes the substance of organic matter that existed in the past. All these things suffered the same fate as we see on Earth.

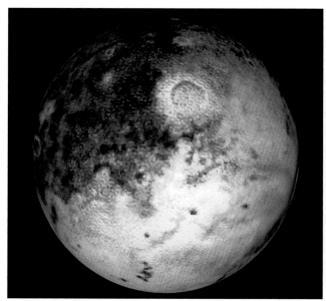 

Raw image   *Image turned upside down*

Image Source: https://static.independent.co.uk

The remains of gigantic creatures abound on Pluto and its largest moon. Where did these beings live in the past… and what purpose did they serve??

It is quite evident that none of these beings and creatures could have inhabited a planet as small as Pluto, or its moons for that matter. So where did they come from, and where did they abode? They may have occupied the space that today we call our solar system, but not as we know it.

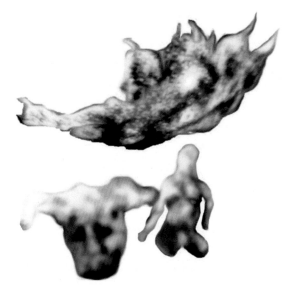

Some frightening creatures are noted…
…while others appear more benevolent.

The photo below is 'the last, best look that anyone will have of Pluto's far side for decades to come', as said by Alan Stern, principal investigator for New Horizons at the Southwest Research Institute in Boulder, Colorado.

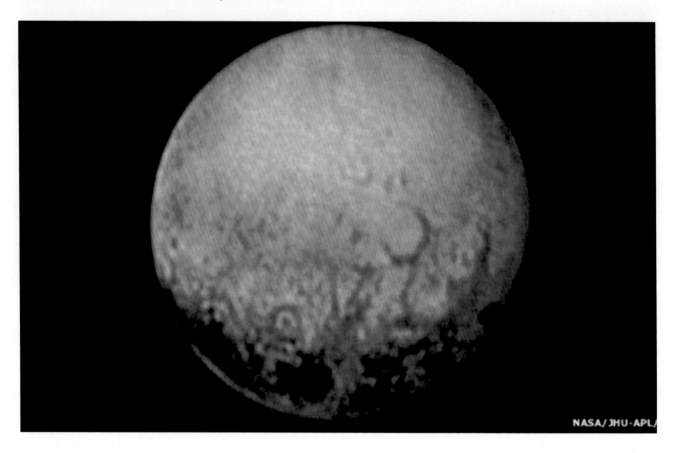

NASA/JHU-APL/

The photo shows a string of four dark spots that have captivated the astronomy community since they were first spotted. In photos taken further away, the spots appeared to be all equally sized. 'It's weird that they're spaced so regularly,' said New Horizons scientist Curt Niebur in a statement.

However, a closer examination covering the bottom half of the Planet produces a shocking view of what actually makes up the rock of this planet.

This is what the filters uncover slightly above the 'dark spots':

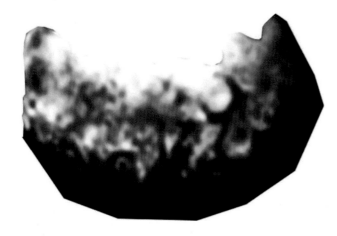 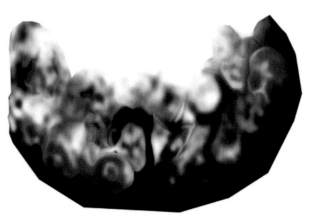

Blur Technique applied only.

Visible features are coloured to distinguish the detail.

Here we have an incredible view of various life forms having lived sometime in the past on a larger planet, possibly in a completely different solar system.

Is humanity ready to accept these facts??

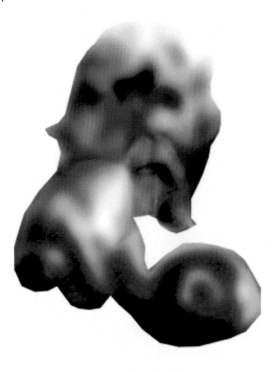

Person with mushrooms?

Person with flowers

The following chapter provides evidence that the entire universe is the substance of organic matter. Here it can be seen that stars became stars only as a result of the organic compound/s that previously existed providing the material for it to become a star. The evidence presented may be considered ludicrous by many, but I believe will be vindicated in the future since all of this will eventually be tested and proven true.

In exposing the 5th Dimension, that which was hidden now becomes revealed as a warning that we have entered this new reality. I believe this to be the Apocalypse, the great revealing….

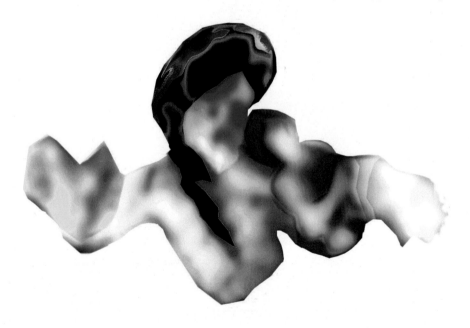

CHAPTER 8

THE UNIVERSE

This chapter is a short preview of further evidence that the entire universe appears as the overarching fractal of the same repeated pattern, namely that all stardust is the substance of previous organic matter, i.e. flora and fauna, that was made from DNA, just like we are.

These life forms proliferated and survived in the same way as we do, made of flesh and blood, and, just like us, also required water, food, clothing and shelter. Many of the things discovered have the same appearance of the things we are familiar with, such as flowers and beings with human-like qualities. In other cases, a range of angels and darker-type entities, including demons and dragons, inhabited these realms in the past, and in vast numbers too.

It is the repeat cycle of the fractal in which we are embedded. When viewing constellations in our own galaxy, we are looking back thousands of light years into the past. An example can be seen in the Westerlund Cluster, a relatively close star cluster at a distance of about 15,000 light years away.

The featured image (below left) of Westerlund 1 was taken by the Hubble Space Telescope towards the southern constellation of Ara (otherwise known as the Altar).

Although presently classified as a 'super' open cluster, Westerlund 1 may evolve according to astronomers into a low-mass globular cluster over the next billion years.

The pictures below demonstrate how easy detail can be identified by utilising only a Gaussian blur technique. This limited detail can then be subjected to other filters, such as adjusting and equalising red, green and blue, and applying contrast and inverting colour values as necessary to uncover the finest of detail, without adding any additional structure to the feature. In other words, it is NOT a drawing.

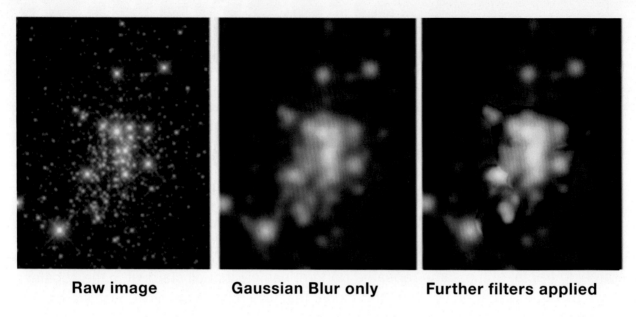

Raw image   **Gaussian Blur only**   **Further filters applied**

Raw Image Credits: https://apod.nasa.gov/apod/image/1706/Westerlund1_hubble_2748.jpg

The eyes are uncovered in exactly the same way. By simply using these techniques it brings out all of the desired features seen in the images presented in this book. Here is an amazing view of a being whose head alone has now become a part of thousands of stars that comprise this constellation. Anyone can utilise the same technique and uncover similar views as I have demonstrated within these pages.

Below is another example showing an exact representation of attributes that we are familiar with as found on Earth today. This confirms the DNA's existence long before we can imagine, including its capability to replicate its code throughout time.

Constellation NGC 2736

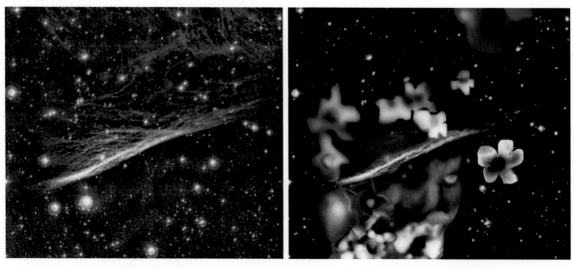

Raw image   **Filters applied**

Credits (raw image): ESO/José Francisco (josefrancisco.org)

Sometimes referred to as the 'Witch's Broom Constellation', this image shows that it was more likely to be the remains of a hat. In this case, we are gazing at a time many thousands of light years ago.

This is light itself, travelling over billions and billions of kilometres still showing the final millisecond when these aspects crystallised and eventually became a part of the stars of our galaxy.

The Unicorn Nebula

Raw image   **Filters applied (Star bug revealed)**

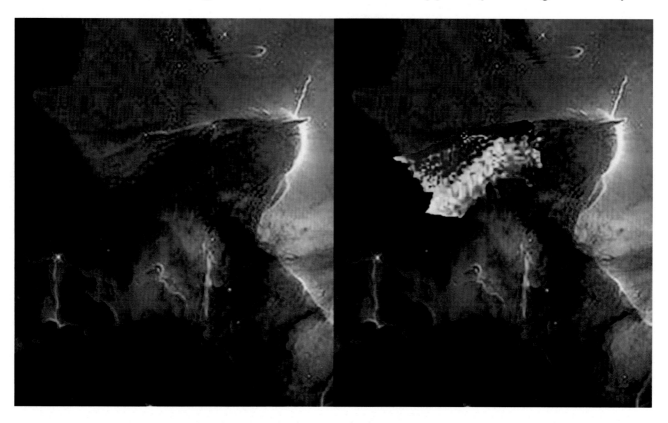

Image Source Credits: https://i2.wp.com/www.rankred.com

An incredible view of an insect so large, it would take many light-years to travel across head to tail.

Star Bug

Who is going to be brave enough, among our cosmologists to attempt to explain, without criticising before doing their own investigations, exactly what is going on over here??

Are we looking at a larger part of the fractal, the very image of aspects that are replicated over time, and getting smaller, each within a fractal space that had similar conditions as we enjoy on Earth today?

Every particle of dust, every rock, planet and star of our galaxy (Milky Way) is the remnant of super-large life forms that once existed, forming our galaxy as we know it. The evidence presented suggests that there was no 'Big Bang'. What other explanation could then be offered...?

Star Man

Rotating the same picture (Unicorn Nebula) counter-clockwise at 45° uncovers more astonishing views of previous generations, going back billions and billions of years, long before galaxies even began forming

Space Dog

Raw Image Source: https://imgc.allpostersimages.com

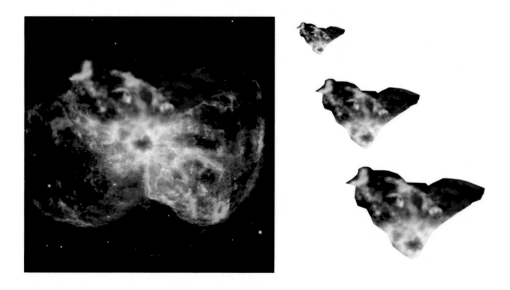

To obtain the detail, the raw image is rotated at 180° with filters applied.

Is there anybody out there who knows something the rest of us mortals have absolutely no idea about? Because if that be the case, then please share this knowledge, and stop pussy-footing around hiding the truth. We deserve to know!

The Pillars of Creation

Raw Image Source: https://steemitimages.com

This 'Nebula' is about 7,500 light years from Earth. I give up trying to understand the dimensions occupied by some of these beings. Note the eyes, forearm, hand and fingers still visible over billions and billions of years. The eyes were uncovered by equalising and inverting the colour values. Other features are also noted.

Although the foreground of this picture portrays some very interesting features, a face of an enormous entity staring seemingly from the background is equally intriguing. Tilting the image 45° to the right gives a better perspective.

Displayed here is the Apocalypse, the revealing of the 5th Dimension, for all of humanity to learn the utter truth of its origins and of the future to come.

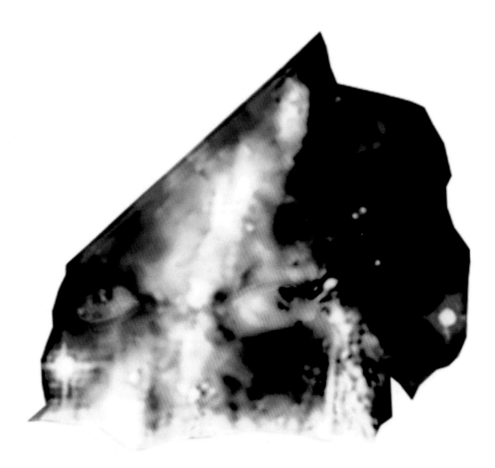

All organic matter, from the very beginning, contained the DNA molecule, an intelligent Code, creating every living thing that ever was and ever will be. The DNA is the beginning of all of creation, and has continued generating life in a never-ending cycle, revealing the eternal nature of the creation.

The Host of Heaven

The image below presents an astounding view of the reality of our Milky Way galaxy, the 'Host of Heaven', comprised of stars of our own galaxy that we had absolutely no idea about; aspects so huge that the beings uncovered seen in the 'Pillars of Creation' would appear as microbes by comparison.

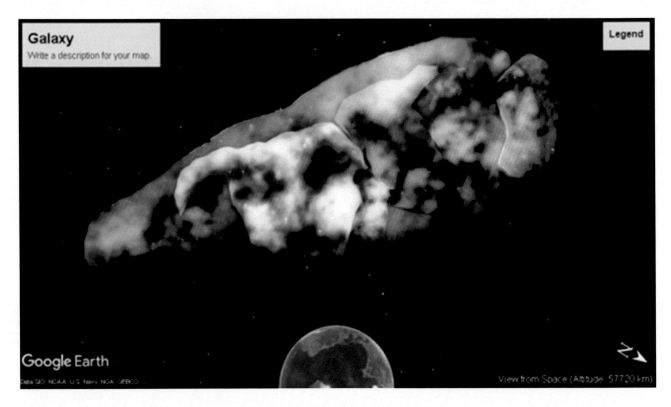

Continuing on from the discussion regarding DNA-containing molecules creating every living thing in existence, both past and present, this means that all other galaxies throughout the entire universe are composed in the same way. In other words, entirely made of stardust derived from previous life forms, some having lived perhaps trillions of years ago based against our timescale.

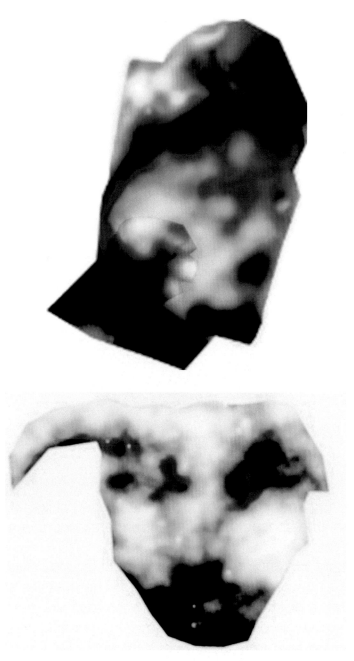

These are visual observations anyone can make by taking the necessary steps to uncover the shocking truth of the existence of entities such as these. This information can NEVER be debunked, as the evidence is there right before our very eyes.

Mr Richard Dawkins... can you please explain this, assuming you will even look?

The revelations in this book defy everything that science has ever taught humanity about the cosmos and its origins. The evidence presented clearly indicates that there was no such thing as a 'Big Bang', as we are looking at things that existed long before this so-called event, which was supposed to have taken place some 13.8 billion years ago.

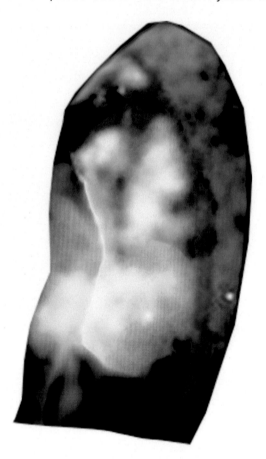

This is calculated from our current understanding of the speed of light. However, these beings and creatures lived in a time-space that defies our understanding in this regard.

To celebrate the imminent 21st anniversary of the Hubble Space Telescope, NASA released the image below of two beautiful interacting galaxies. The massive spiral galaxy, which is known as UGC 1810, has a disk that is distorted into the rose-like shape by the gravitational tidal pull of its partner galaxy, known as UGC 1813. The beautiful blue points along the top are comprised of combined light from clusters of incredibly bright and hot young blue stars.

Raw image

Image rotated 90° counter-clockwise with contrast added to reveal detail.

Serpent–like creature

The image on the right reveals an astonishing view of a snake's head, making up UGC 1813. This is in the constellation Andromeda, around 300 million light years from Earth.

Although the galaxies appear close, they are tens of thousands of light years from each other apart from a thin bridge of stars, which are probably part of the rest of the serpent.

Here we observe further evidence that galaxies themselves came into existence following what existed before, as in the various life forms making up the matter that now forms our known universe. The speed of light must have been relative to the dimension of space-time they occupied, which is why we cannot estimate how far back in time they could have existed.

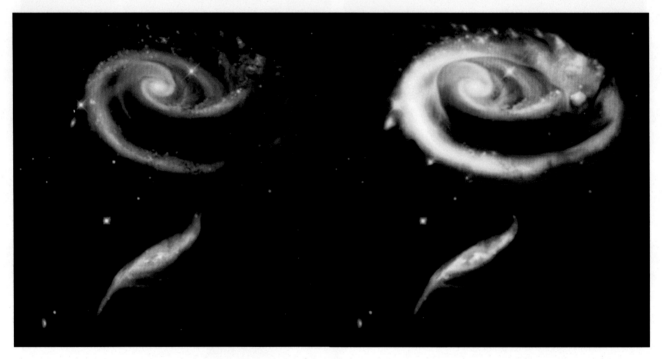

Raw image   Uncovering the Dragon

Furthermore, this image also reveals what can only be a dragon. Keep in mind that we are gazing at a galaxy said to be 300 million light years from Earth, meaning the light from this galaxy has taken 300 million years to reach our planet.

We are looking back to the past when this dragon had already crystallised and begun forming this galaxy. However, the question is: when and where did it reside, together with the snake, before crystallising and eventually becoming the stars that we currently see forming this distant galaxy more than 300 million years ago?

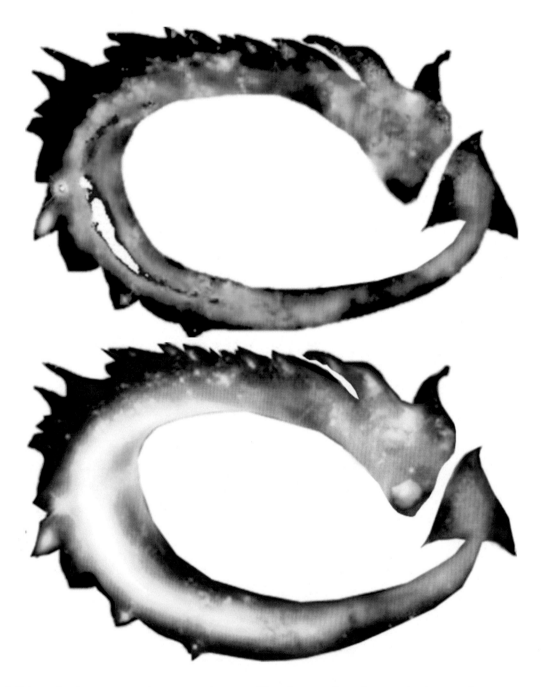

Other striking features can be noted, specifically on the dragon's face when inverting the colour values.

For the open minded, this shows how our past and our future are nested in a much greater scheme. We are observing evidence of the existence of the DNA Code (which I believe to be the 'Word of God', as mentioned in the Bible) when it all began.

This Code determines every beginning and every ending, the Alpha and Omega. Our thoughts and actions alter elements within the Code, moulding us into what we become, adjusting our level of consciousness accordingly. We become the very things we believe and think about.

This is far more profound than what any human has ever imagined, especially concerning the creation of the universe. Here presented is tangible evidence that we reside in a fractal (the 5th Dimension) and that the Big Bang theory quite simply isn't true. Instead, what existed from the beginning was DNA… the Code (Word of God) from which all living things (organic matter) originate.

The original organic matter became dust and rock… and thus began the cycle; but as a reduction, not an expansion as the Big Bang theory suggests. In the beginning, everything started off bigger than the universe itself, and all organic matter was then (and still is) reduced over time by means of an unknown event taking place in cycles, based on the Fibonacci sequence. Yes indeed, we are the dust of the stars and our consciousness is influenced directly from what existed in the past, including what will exist in the future – all derived from **The Rock of Ages**, the **Tree of Life**.

In every respect, the things of the past reveal the exact truth of our origins, which is probably more akin to mythology than what we would ever imagine.

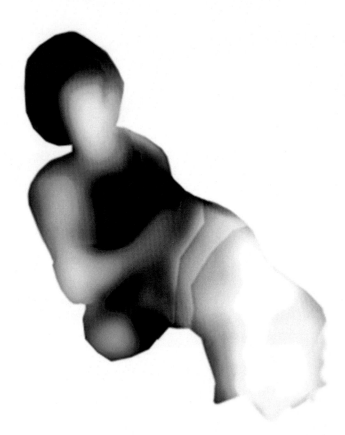

This is the 5th Dimension, the Age of Aquarius, the Water Bearer.

CHAPTER 9

A NEW BEGINNING

In the beginning was the Word (DNA Code), and the Word (Code) was with God, and the Word (Code) was God. The same was in the beginning with God. All things were made by Him.

John 1: 1 (KJV)

Every Beginning has an Ending, and every Ending has a new Beginning.

Based on the evidence presented within these pages, it would seem that many creations may have taken place – perhaps even billions. Certainly, some of these beings would measure thousands, if not hundreds of thousands of light years across the span of their bodies. We are clearly created out of the dust of previous creations and the dust of our current cycle will become the substance for the next creation. This I believe is the mechanical process of the universe based on an intelligently-designed fractal, revealing the eternal nature of creation and its Creator.

So, do I have all the answers…? To be exact, absolutely not!!

However, a **five-dimensional space** is a space with five dimensions. If interpreted physically, this means it is one more than the usual three spatial dimensions and the fourth dimension of time used in relativistic physics. It is an abstraction which occurs frequently in mathematics, where it is a legitimate hypothesis.

Whether or not the actual universe in which we live is five-dimensional is a topic of debate. However, in light of these observations I would strongly suggest that the 5th Dimension is the fractal itself in which all of existence will ever be. Within the greater picture, our own time-space is embedded, playing out a similar role of aspects that existed before.

Many real-world phenomena exhibit limited or statistical fractal properties and fractal dimensions that have been estimated from sampled data using computer-based fractal analysis techniques. Practically, in its current application, measurements of fractal dimensions are affected by various methodological issues and are sensitive to numerical or experimental noise and limitations in the amount of data. Nonetheless, the field is rapidly growing as far as estimating fractal dimensions for statistically self-similar phenomena is concerned and may have many practical applications in various fields of science.

This book demonstrates that the entire cosmos consists of the remains of organic matter, forming the dust making up the stars and the planets that we see in our universe today. The scale of the fractal is revealed beyond the Milky Way and possibly extends beyond the galaxies themselves.

The planets are made up of smaller life forms, forming the geology they are made of. This is how the entire universe is structured: out of previous ***biological matter***. Does this then mean that matter existed before energy? And if so, *where does this place the equation $E=MC^2$…?*

However, the fractal does not explain the nature of how it all got to be like this. The only explanation that I can offer is that of a Photon Belt / superwave, which itself is governed within the fractal. This appears to have been a repetition over billions of eons.

Below is a catalogue of the various entities found on Earth, as displayed in this book, listing their estimated standing heights, calculated by measuring head dimensions using the Google Earth ruler. Once the length of the head is established, the estimated standing height of the being can then be scaled proportionately.

Most of these, I am sure, could never have coexisted together in the same dimension of space and time as we know it. For example, it would seem impossible for the Dark Emissary and the Sentinel to have coexisted in the same space-time dimension. By comparison, this could be compared to an average man today having to compete with 14-metre giants.

The list is arranged in sequence, starting with the biggest beings found on Earth thus far:

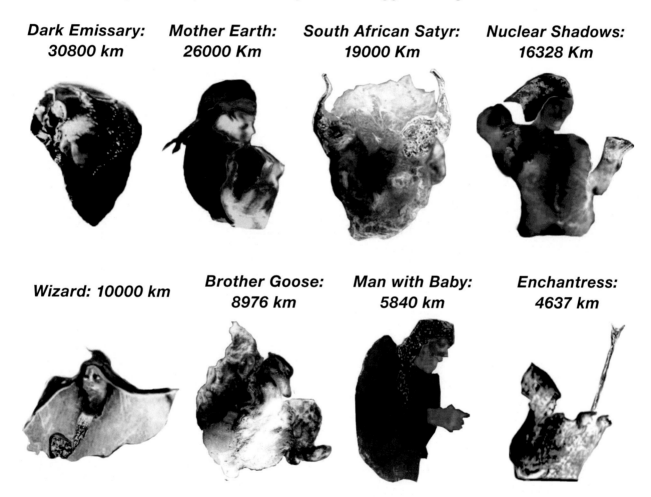

| **Dark Emissary:** 30800 km | **Mother Earth:** 26000 Km | **South African Satyr:** 19000 Km | **Nuclear Shadows:** 16328 Km |
| **Wizard: 10000 km** | **Brother Goose:** 8976 km | **Man with Baby:** 5840 km | **Enchantress:** 4637 km |

Sentinel: 3867 Km

Priest with Scroll: 2300 Km

Dragon Rider: 1900 Km

Soothsayer: 1900 km

Beauty & Beast: 1838 km

Sabrina: 748 km

Warrior Monk: 640 km

Bedouin: 450 km

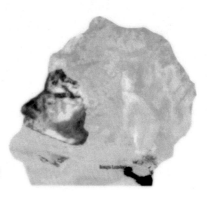

Moses: 320 km

Sorcerer: 205 km

Shaman:187 km

Sinai Reaper: 66 km

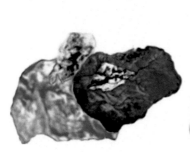

Sinai Merchant: 55 km

Beautiful Amazon: 40 km

Egyptian Princess: 28 Km

Ancient Angel: 21Km

Antarctic Priest: 7.3 km

Sage: 6.5 km

Beautiful Angel: 5.1km

Antarctic Ghoul: 3.66 km

Angel with Scroll: 1.64 km

Ice Angel: 840 meters

Merlin: 135 meters

Muscle Man: 41 meters

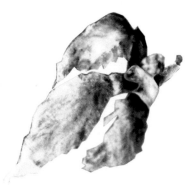

Measurements were further analysed in terms of comparing the size ratio difference between the biggest entity matched in sequence to the next biggest sample, which could be measured with some degree of accuracy. Starting with the Dark Emissary and working back to the Ice Angel produces a trend that appears to be cyclical.

What is interesting are the significant changes which appear as peaks and valleys forming a rather regular pattern. If we are to accept a theory based on different sizes = different space-time cycles, then this could be possible evidence showing cycles of reduction occurring over time. The graph below, titled 'Ratios Measured in Size Order' shows how certain entities, when paired in sequence (bigger to smaller) may reveal possible cycles, specifically where the ratio is greater than or equal to 1.5. A ratio of 1.5 means that the previous life form was at least 50% bigger against the one being compared.

The regular pattern that is noticed could possibly depict cycles of time when these life forms could have or could not have coexisted during the same time interval.

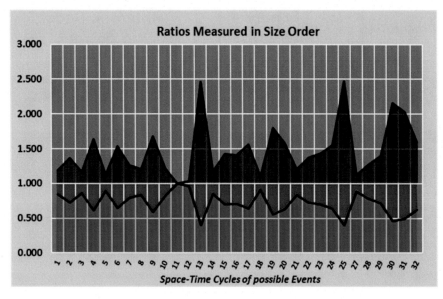

If we assume a starting point going back 4.6 billion years, which is the estimated age of the solar system based on the time period given by cosmologists, then the two largest beings found on Earth thus far, i.e. the 'Dark Emissary' and 'Mother Earth' must have already existed long before this time, since they must have occupied a planet with a much larger surface area with a much greater atmosphere.

All things were suitably proportionate to support them during that cycle.

The starting point on the graph (1) is the ratio between the Dark Emissary and Mother Earth, producing a value of 1.185, matching each other by about 84% in height. This means that they could have coexisted in the same space-time dimension. Note also that females are generally slightly shorter than males. On the other hand, the ratio between Beauty and the Beast and Sabrina is 2.457. This means that Sabrina is about one and a half times smaller by comparison. The obviously similar-sized life forms, such as Dragon Rider, Soothsayer and Beauty and the Beast are possible candidates that may have existed in the same space-time frame.

It must be noted that these assumptions can only be made regarding the acceptance of the theory based on **different sizes** = **different space-time cycles**.

Referencing the actual time period may be impossible due to our current understanding and teachings regarding the formation of the solar system and the speed of light. It is clearly evident that most of the beings catalogued in this book could not have survived on Earth, nor in the solar system as we know it.

Does the evidence presented suggest that there could be sudden and seemingly cataclysmic endings followed by new beginnings? And if so, does each new beginning then assume a smaller physical dimension of what existed before?

This indeed appears to be the reality of the cosmos, its entire structure based on the Golden Ratio, Phi, 1.618, sometimes referred to as the Fibonacci sequence.

The above is demonstrated through the measurements obtained from the various samples observed, which by comparison show a significant probability that the differences match the Golden Ratio.

Further evidence supporting the idea of a fractal is demonstrated in this analysis. Here, **actual** measurements are compared against **predicted outcomes** calculated against the Golden Ratio.

The comparison produces a correlation coefficient of 0.983. This demonstrates how the measured data exhibits a significant relationship aligned to the Golden Ratio – 1.618.

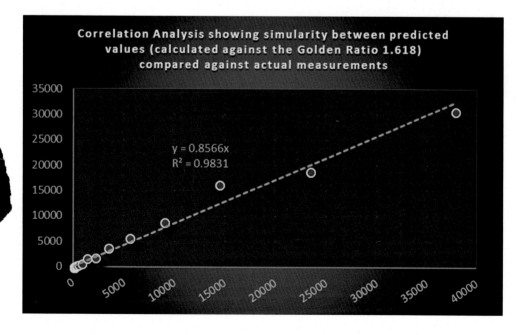

Correlation Analysis showing simularity between predicted values (calculated against the Golden Ratio 1.618) compared against actual measurements

$y = 0.8566x$
$R^2 = 0.9831$

This significant result could possibly provide much supporting evidence that we are in fact part of a fractal arrangement, determined by the Golden Ratio.

This itself occurs in cycles, possibly continuing without ever ending, depicting the everlasting nature of the Creation.

Science and religion are challenged to explore the evidence presented in this book to help us understand the cycle of organic matter.

Now is the time for both institutions to put aside petty arguments about the creation and rather provide a possible answer and understanding to what these observations might mean.

The information in this book is way beyond what most of humanity can accept. Chewing on a

revelation of this magnitude without choking will be a challenge to most people. Only a small percentage of the population who see this are likely to accept it as fact.

Unfortunately, this information may never be published in the mainstream media because it is controlled by higher powers who do not want you to be educated with the truth.

The longer everyone remains comfortable with the status quo, the longer we will remain ignorant of the truth. And ignorance comes with a price.

I believe this is the Great Revealing – the Apocalypse we have been expecting. We can now see our origins stemming from the **Rock of Ages**.

These discoveries are a direct result of the coming of the Age of Aquarius, the 5ᵗʰ Dimension.

It cannot be avoided; it is a cycle into which many will enter a new state of consciousness and perhaps witness the next event. It will be sudden and unexpected; and when it happens…

….it will bring a **New Beginning**.

ACKNOWLEDGEMENTS

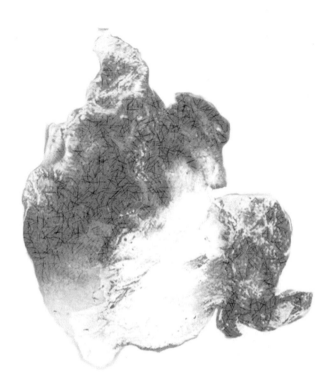

- Google Earth.
- 2015 DigitalGlobe Image Landsat.
- Data SIO, NOAA, US Navy NGA, GEBCO Image Landsat, Image IBCAO.
- 2015 CNES / Astrium Image Landsat.
- 2015 CNES / Spot Image
- 2012 Encyclopaedia Britannica, Inc.
- Robert Stanley, The Photon Belt Part 1 & 2.
- NASA (http://mars.nasa.gov/multimedia/interactives/billionpixel/)
- Zachariah Sitchin, the Lost Book of ENKI.
- The Fifth Dimension – Aquarius
- Wikipedia
- US Geological Survey
- NASA's New Horizons Probe, Pluto, and Charon
- Hubble Space Telescope

Statement from the Author regarding the Fair Use and Attributions to Content regarding Google Earth Copyrights as Modified on December 17, 2015:

No 'content' has been included in the images, as all these features have been deactivated as per Google Earth settings to exclude things such as place names, boundaries, cities etc. All images presented are directly available '***online***' and are classified as '***raw images***' wherever they are presented in the book. All other images are considered '***offline***', since these are not how the images would appear '***online***' and can only appear in the detail as attributed by utilising suitable software applications.

Declaration from the Author

I compiled this book out of desperation, seeking answers to an extraordinary discovery I made in early 2013. I happened to make some observations that I believe will test both the scientific and religious communities. This book is a product of a chance discovery leading me to compile and publish these observations (as images), including a personal interpretation of what these observations might mean. Having no formal qualification or background in the fields of cosmology, astronomy, physics, palaeontology or archaeology, it is nevertheless obvious that the evidence derived from this discovery will challenge the institutions teaching these disciplines. Theologians are not excluded.

My hope is that other interested parties will engage on a grand scale utilising a similar modus operandi, as I have, to uncover comparable aspects en masse. This will negate any attempt to debunk this information. I declare all the observations published are real, albeit that each of the characters named may not specifically have played the role suggested in the title identifying them.

I'm a South African, born in 1954 and grew up in a mining town situated on the East Rand, approximately 45 kilometres from Johannesburg. My background is that of a SHEQ management specialist and consultant, specialising in integrated management systems as well as auditing management systems for certification purposes, in compliance with international standards such as ISO 9001, ISO 14001 and ISO 45001. I am a certified quality engineer and a qualified auditor.

If you would like to contact me about anything concerning the book, I would be happy to hear from you.

My email address is lionelingram1954@gmail.com

Lionel Ingram
Author

Printed in the United States
By Bookmasters